THE MAKING OF VICTORIAN
SEXUAL ATTITUDES

The Making of

VICTORIAN SEXUAL ATTITUDES

MICHAEL MASON

Oxford   New York

OXFORD UNIVERSITY PRESS

1994

Oxford University Press, Walton Street, Oxford OX2 6DP

Oxford New York
Athens Auckland Bangkok Bombay
Calcutta Cape Town Dar es Salaam Delhi
Florence Hong Kong Istanbul Karachi
Kuala Lumpur Madras Madrid Melbourne
Mexico City Nairobi Paris Singapore
Taipei Tokyo Toronto
and associated companies in
Berlin Ibadan

Oxford is a trade mark of Oxford University Press

Published in the United States
by Oxford University Press Inc., New York

British Library Cataloguing in Publication Data
Data available

Library of Congress Cataloging in Publication Data
Mason, Michael, 1941–
The making of Victorian sexual attitudes / Michael Mason.
Includes bibliographical references (p.   ).
1. Sexual ethics—Great Britain—History—19th century.
2. Great Britain—Moral conditions.   I. Title.
HQ32.M36   1994   306.7′0942—dc20   94–16822
ISBN 0–19–812292–6

Set by Hope Services (Abingdon) Ltd.
Printed in Great Britain
on acid-free paper by
Bookcraft Ltd.,
Midsomer Norton, Bath

For my family

Preface

This book started life as the second part of a one-volume study of nineteenth-century English sexuality which grew to an unwieldy length. I have published the two parts as separate books. As it happened, the project divided naturally around the halfway point, so that the analysis of actual behaviour and of what may be called primary beliefs about this behaviour—doctrines in nineteenth-century medicine and sociology—is all dealt with in my first volume, entitled *The Making of Victorian Sexuality*. The present volume is concerned with beliefs of the day which were only secondarily about sex: beliefs about religion, general morality, and politics. It seeks, in fact, to bring out the place of sexuality within some of the period's major ideological schemes.

I have only revised these chapters slightly for publication in their new setting, as I believe they stand up as a separate contribution to the understanding of sex in nineteenth-century England, without reference to my previous discussion. However, it is probably worth stating here the gist of *The Making of Victorian Sexuality*—which was not to deny the anti-sensualism popularly attributed to the 'Victorians', but rather in one way to assert it more emphatically, since it was found to be a widely and warmly embraced creed. This very shift of emphasis creates the need for the present inquiry about ideological origins, because the usual ideas about the latter become inadequate. Nineteenth-century anti-sensualism could not have been driven by the fairly puny forces which are normally invoked: neither by the lip-service puritanism of the bourgeoisie, nor by the asceticism of Evangelicals. It simply had too much vitality for that.

Contents

❧

List of Illustrations

Chapter One

'THE THEOLOGICAL VIEW'

I

Higher Natures

AUSTIN HARRISON'S account of his celebrated father Frederic, published in 1926, is an early example of a book embodying a fully fledged notion of the 'Victorian' in the modern sense of a disparaged era of sexual moralism. It is also the last in the important line of English nineteenth-century father-and-son memoirs which runs from that of Francis Place, through those of John Stuart Mill and Edmund Gosse, memoirs which use the idea of male generations to enact moral-historical transitions.[1] Frederic Harrison emerges from his son's account as an exceptionally lovable and rewarding father, and also as an utterly 'Victorian' man. The conjunction is intentional: Austin Harrison wants to portray an individual whose extreme moral strictness was not forced or hypocritical, and not a source of discomfort or smugness. No other single document, perhaps, is so effective in reminding us that the 'Victorian' Victorians were not all Chadbands, Gradgrinds, and Podsnaps, but very commonly human beings like ourselves, comfortable with, and even dedicated to, a moral code which—unless we understand its roots—might strike us as one that only a monster or a madman could espouse.

This is not to say that Austin Harrison accepted his father's Victorianism. On the contrary, he felt deeply sundered from it, especially on the matter of sexuality. Frederic, born in 1831, had

---

[1] And for a working-class example see also Wilson 1909.

remained chaste until his marriage at the age of 38. Though a friend
of George Eliot, he seems to have regretted her liaison with George
Henry Lewes.[2] His son on one occasion 'determined to "sound" him
on the question of social morality', and he put the question, 'what is a
fellow to do who cannot marry and falls in love?' For Frederic the
answer was 'morality': 'a man who cannot learn self-control is a cad
. . . A loose man is a foul man . . . anti-social . . . He is a beast'. And he
agreed with Austin that positivism—the humanist creed which he
had adhered to for most of his life—'takes the theological view about
morality'.[3]

The notion that apparently divergent forces in English nineteenth-
century culture—on the one hand religious and on the other secularist—actually converged and reinforced each other in certain
important areas is attractive. In a broad form this idea has been
famously stated by Elie Halévy, who made Evangelicalism and
Utilitarianism the formative agents in early Victorian society. In the
sexual domain it was observed by W. H. Mallock as early as 1891 that
both free-thinkers and religious groups attached a significance to
adultery (he was writing about the Parnell affair) which was inconsistent with their other principles, and must flow from some third doctrine which they shared. Mallock identified the latter as an
'instinctive utilitarianism' which had become 'clenched' and 'rigid'.
More recently Muriel Jaeger has written of how the 'strong rationalist element that persisted . . . throughout the Victorian Age . . .
shared the Puritanical austerity of Victorian religion'.[4]

On the other hand, in thirty years of controversy over the
Malthusian doctrine of 'moral restraint' in sexual intercourse, the
most anti-sensual views were not voiced by the 'theological' participants (as I have shown in the first part of this study). Militant
Christian opinion on the nonconformist and Low Church wing took
against Malthusian sexual self-control more than any other sector,
and if this pattern persisted in the Victorian years it would be very
damaging for the conventional idea that Victorian sexual moralism
was particularly promoted by lower church and dissenting groups
among Christians of the period. Of course Frederic Harrison, who
was an expert observer of the sects of his day, may not have had this
brand of opinion in his sights when he told his son that positivism
took the 'theological' view on sex. He may not even have felt the con-

---

[2] Vogeler 1984: 57.   [3] A. Harrison 1926: 127–8.
[4] Mallock 1891; Jaeger 1956: 103–4.

vergence of his creed with religious doctrine, in this instance, to be especially paradoxical. For Harrison's positivism was a complicated and elusive bundle of beliefs such as can often be encountered in the broadly 'secularist' zone of Victorian thought. Its foundation was the teaching of Auguste Comte, whose progressive and humanist system led Harrison to an extreme, almost Communist, radicalism in politics, and to the quasi-religious Religion of Humanity. Ontologically, the latter was meant to be empty of supernatural content, but it had its prescribed rituals, and Harrison seems always to have ducked metaphysical questions, claiming that they were unanswerable and unimportant.

Nevertheless, while remaining conscious of the many nuances in nineteenth-century religious belief and disbelief, I shall suggest in the following chapters that explicit anti-sensual attitudes tended above all to emanate from secularist and progressive quarters: from individuals trying to apply the thought of Jeremy Bentham, for instance, certainly from those who might write for the *Westminster Review* and the *Fortnightly*, equally certainly from women's rights activists, and even from those who would be found addressing Chartist rallies. Church and chapel of course contributed to anti-sensualism, but probably chiefly in a passive fashion, through the traditional moralism of mainstream Church of England opinion. It does not seem likely that Anglican Evangelicals gave much of a boost to anti-sensualism (they may even have exerted an opposite influence). Nonconformist Evangelicalism, especially in Methodism and in those groups committed to a kind of social engineering, has stronger claims of this sort, but even here the picture is ambiguous.

In another anecdote of the period, the 'theological view' about sex actually emerges as more lenient than that of one of the great leaders of progressive opinion.[5] William Johnson Fox was a public voice of consequence in the early 1830s, since he combined the office of minister to the Unitarian South Place Chapel with the editorship of an important magazine, the *Monthly Repository*. His public and private existences merged suddenly in 1834 when members of his congregation, acting on complaints from Mrs Fox, reproached him for his conduct with his young ward, the 30-year-old Eliza Flower. This had been giving offence to his wife for some time: the Foxes had as a result been living separately in the matrimonial home since 1832.

[5] On the Fox–Eliza Flower events see Garnett 1910: 155–6; Mineka 1944: 188–97; Johnson 1979: 31–2; Budd 1977: 19–20; Rover 1970: 31–7.

Fox twice challenged the loyalty of his congregation by resigning on the issue, and was twice reinstated, though on the second occasion about a quarter of the members seceded in protest.

Fox denied that he and Eliza, who were certainly in love with each other, had had intercourse. But it does not much matter if this denial was truthful, since, interestingly, Fox did not offer this supposed abstention by the couple, or abstention by any couple in such circumstances, as meritorious. The loyalists in his congregation seem to have granted the point, since following their endorsement Fox was at once able to leave his wife and set up house with Eliza. In his own case Fox always explained that only circumstances had made consummation of the relationship inexpedient, and through the *Monthly Repository*, at this date, he lent his support to divorce and even adultery as general principles. In successive issues of the magazine in 1833 he suggested that 'polygamy' would be preferable to prostitution on its current scale (he evidently assumed that both prostitutes and their clients were in many cases parties in unhappy marriages), and that Mehetabel Wesley, sister of John and Charles Wesley, would have been justified in leaving the husband she had been forced to marry in order to live with the man she loved: 'then she would have been liberated from an enforced and intolerable bond, and even have entered on a new state, perchance, of the affection and enjoyment for which she was framed'.[6]

The most eye-catching contribution to the *Monthly Repository* in this vein was by William Bridges Adams, one-time son-in-law of Francis Place, writing as 'Junius Redivivus'. His article 'On the Condition of Women in England' is a plea for equality and freedom in marriage, with the right to divorce by mutual consent. This policy, exceptionally advanced for its time and a cause of scandal to some readers of the *Repository*, is grounded in a bleak picture of modern marriage which Bridges insists on with vehemence, even with a kind of gloomy relish: at all levels of society are legions of miserable couples living out the hangover of rash marriages contracted under the influence of greed or lust. Though the terms of Bridges's diagnosis and cure are extreme his anxiety about loveless courtship and marriage is shared by a surprising range of commentators in the 1830s and 1840s, from socialist Owenites to conservative Mrs Sarah Ellis,

---

[6] W. J. Fox 1833*a*; 1833*b*.

and there seems to have been some sense at the time that these institutions were in crisis.[7]

John Stuart Mill had close links with Fox: Mill wrote for the *Repository* and Fox for the *Westminster Review*, while Harriet Taylor—with whom Mill had just embarked on a platonic affair—and her husband John were important members of Fox's South Place congregation. Above all, Eliza Flower was Harriet's closest friend. Mill, speaking also it seems for the Taylors, wrote a remarkably serpentine letter of advice to Fox in July 1834:

We all most decidedly think that since the crisis in the congregation appears to have been brought on principally by the belief that a fact, which would be of the greatest importance in their eyes, though of none at all in yours, is true—it would be very foolish that you should not have the full advantage of its not being true . . . While that fact is denied and deniable, all who are otherwise favourably disposed will not be afraid to stand by you . . . But if it were made impossible for anyone to defend you except those who were willing to encounter the odium of justifying all which is now alleged against you, I am afraid . . . people will make it a matter of conscience to discountenance what they consider the open profession and vindication of immorality.

This being the case, I should not, if I were in your situation, think myself bound to court attention to the fact that expediency only and not principle was the cause of your not having gone the full length of what they assert. If they put that question to you, no doubt you ought to say so—but I think not otherwise. It seems to me quite enough if you appeal to those articles in the Repository as containing your principles on the subject. You might say that you acted no otherwise than in consistency with those principles; and if they ask you whether the particular fact is true, you might deny altogether their concern with it or right to enquire into it, but nevertheless profess your willingness voluntarily to give this information sought, by denying the assertion. We all think it of great importance that every public mention of the charge should be accompanied by mention of your denying it—and also that the effect of this denial should not, unless it be absolutely necessary to your integrity, be inspired by the public profession of the extent to which your principles go in that one matter.[8]

This combines a plan for the manipulation of the public response to the affair, not unworthy of President Nixon in 1973, with a veiled threat to withdraw support unless Fox put his practice and theory on adultery in the most moral light consistent with the views he had already stated in the *Repository* articles. Overtly it is 'they', 'people'

---

[7] W. B. Adams 1833; see also O'Donnoghue 1836: *passim*; Ryan 1831: *passim*.
[8] J. S. Mill 1963: 228–9.

of a moralistic cast, whose prejudices have to be appeased by a persistent denial of the 'fact' of intercourse with Eliza, but this group
shades implicitly into the trio of friends giving advice. Such projecting of supposed public disapproval is, as we shall see in Chapter 3, a
classic device of Victorian progressive anti-sensualism. The damage
that may be done to the cause of progress by prudish public opinion
on sexual matters is a pretext for assenting to its moralism, but it is
also notable that a need to be roundabout in one's anti-sensualism is
perceived in the first instance. In fact, Mill (and presumably the
Taylors) tend to acknowledge that there is something uncomfortable
about being as hostile to adultery as less enlightened people, and
they probably had some genuine ambivalence on the matter: they
are, after all, giving friendly advice to a man for whom, as they recognize, the technical fact of sexual intercourse was a shibboleth.

Significantly, Mill and his friends overestimated the dangers of an
imprudent frankness. Fox did not take their advice. He did not help
his congregation to believe that his refraining from intercourse with
Eliza had been principled and willing. And he carried most of them
with him. Personal loyalty, however, must have had much to do with
this and, even at the level of ideology, it may be argued that it was
not as Unitarians that these well-educated middle-class Londoners
gave their liberationist minister their backing.[9] Not only did the
other Unitarian chapels in London all disassociate themselves from
South Place, but Fox himself led his surviving congregation in a candidly secular direction after 1834, dropping almost all ecclesiastical
features of its practice and terminology. He was no longer the
'Reverend' William Johnson Fox, and he gave social and political
speeches to an audience no longer designated a 'chapel'—which was
later to evolve into the South Place Ethical Society, whose officials
felt able, by the 1870s, to sponsor free-love policies.[10]

Theological authority in the Unitarian movement was opposed to
Fox simply on the strength of his views on divorce, in the abstract.
The Reverend Lant Carpenter had already protested in a letter to
Fox in 1833 about the tenor of the 'Junius Redivivus' piece—in other
words, as he complained when *The Times* attributed Fox's position
on divorce to the Unitarians as a whole, before he had 'the slightest
idea of the degree in which the purport of it was applicable'. The
commanding Liverpool Unitarian James Martineau, while deploring

---

[9] But on this see also Rossi 1970.   [10] Coupland 1878: *passim*.

the spirit of Fox's enemies, and hoping that he would remain within the national fold, found his opinions on marriage 'obnoxious', 'ethically unsound', and 'mischievous'.[11]

This core of conservative moral opinion in the Unitarian church did not derive, however, from moralistic Christian belief. This would not have been in keeping with the progressive and heterodox character of the Unitarian attitude, various though its expressions were. The church with which John Stuart Mill was linked, via South Place, had a strong tradition of rationalistic anti-sensualism. The key authority here, often cited in the nineteenth century, was Joseph Priestley's *Considerations for the Use of Young Men*, first published in 1776. Above all, Priestley gives religious sanction, of a dry and intellectual sort, to the belief that can be seen gathering momentum in the thirty years of Malthusian controversy, and which was so in tune with Victorian environmental moralism, to the effect that sexual drives are very much subject to cultural influences. The 'plea . . . often used for carnal indulgence' that 'the constitution required it' is, according to Priestley, quite false: on the contrary, 'the mutual inclination of the sexes, if it was not accelerated by an improper conduct of the mind' could be delayed in its emergence, and prostitution would not have developed if a population of prostitutes had not been created by the practice of 'debauching a young woman' (this will become an important theme in the next century).[12]

So Unitarianism illustrates the breadth of sexual-moral doctrine which could fall under a Christian rubric—a point I shall return to shortly—even to the extent of participating in the sort of instability on sexual codes which is characteristic of certain purely secularist movements in the nineteenth century. A witty commentator of 1810 put the case that the anti-sensualism of Priestley and the libertarian spirit of the German dramatist Kotzebue sprang from the same assumptions:

Priestley, as well as Kotzebue, assumes the principle, that both sexes have like rights, and like duties. Kotzebue infers from this principle, that women are to practice a masculine morality, and to indulge in promiscuous intercourse. Priestley infers from this principle, that men are to practice a female morality, and to have no sexual intercourse before matrimony.

[11] *Christian Reformer*, 4 (1837), 234–6; *Monthly Repository*, 7 (1833), 347–57; Garnett 1910: 169–70.
[12] Priestley 1791a: 4–5.

When William Johnson Fox comments derisively in 1834 on the overriding value attached to chastity as a virtue in women he is speaking from the edge of Unitarianism as a denomination, but not from beyond it as doctrine. And in 1837 he still looked back on his 1833 statements on marriage and divorce as 'amongst the last of many efforts . . . to arouse them [Unitarians] to what I thought their proper position in the world as religious and social reformers'.[13]

As I have mentioned, Mill's letter is an equivocal document: setting its face firmly against sexual laxity, but perhaps not at ease with the alignments which are entailed. Over his whole career Mill certainly displayed ambivalence on sexual morality, for he is known to have been active as a young man in Francis Place's campaign for device-based birth control, though later in life he rejected such policies and suppressed the fact of his involvement with Place. This episode is discussed in Chapter 4. Mill's father James also appears to have favoured contraception as a solution to population problems, but in terms of public utterance he did not go beyond some veiled remarks in an *Encyclopaedia Britannica* article on emigration. There is a brief but important summary of James Mill's views on sexuality in his son's *Autobiography*:

In ethics . . . he was supremely indifferent in opinion (though his indifference did not show itself in personal conduct) to all those doctrines of the common morality, which he thought had no foundation but in asceticism and priestcraft. He looked forward, for example, to a considerable increase of freedom in the relations between the sexes, though without pretending to define exactly what would be, or ought to be, the precise conditions of that freedom. This opinion was connected in him with no sensuality either of a theoretical or of a practical kind. He anticipated, on the contrary, as one of the beneficial effects of increased freedom, that the imagination would no longer dwell upon the physical relation and its adjuncts, and swell this into one of the principal objects of life; a perversion of the imagination and feelings, which he regarded as one of the deepest seated and most pervading evils in the human mind.

In the troubling distinction between James Mill's 'supreme' theoretical distaste for conventional morality but practical conformity to it, and in the admission that he had no precise idea of what a liberated sexual system would be like, we are not far from the convolutions of John Stuart's advice to Fox. James Mill's very Godwinian prediction about the future course of human sex-drives was perhaps a

<hr>

[13] *Monthly Magazine*, 30 (1830), 49; W. J. Fox 1834; *Christian Reformer*, 4 (1837), 282.

means of reconciling progressive attitudes with a disgust for sexuality. As the passage unfolds there does emerge, despite his son's best efforts, the picture of a man with a neurotic aversion to sex. (John Stuart deleted all references to his mother from the published form of the *Autobiography*. In the manuscript it is said that there was no warmth between wife and husband, although the latter—for whom she spent her life 'drudging'—fathered ten children on her.) In his anti-sensualism James Mill seems to have gone quite as far as, if he did not surpass, any known nineteenth-century advocate of the 'common morality'.[14]

This was all probably a considerable modification of the views of the master, Jeremy Bentham, though there is no record of Bentham's thinking on these particular questions. He regarded heterosexual marriage for life as the most beneficial and most natural kind of sexual union, but granted the legitimacy of sexual pleasure *per se*, in any setting. He defended concubinage as an expedient in an unreformed society, and had an admirable perception of the cruel and sinister character of the legal treatment of homosexuals. There is in fact a thin line of libertarian comment in the nineteenth century which finds support in Benthamite consequentialism, or the greatest happiness argument.[15] But if the direction in which James Mill drifted from canonical Benthamism on sex may be called Godwinian his son's distinctive heresies are perhaps better described as Shelleyan. John Stuart did inherit the Godwinian view on sexual drives, and he remained faithful to it. 'I think it most possible', he wrote in 1870, 'that this particular passion will become with men . . . completely under the control of reason'.[16] But there was a Shelleyan supplement, which accords with Mill's own well-known analysis, in the *Autobiography*, of his mentality as that of Benthamism enlarged and softened by contact with the arts, and especially with poetry.

The best evidence on the matter is the remarkable twin document 'On Marriage',[17] in which Mill and Harriet Taylor each made a statement, for the other's eyes, of sexual-moral beliefs. This probably dates from the time when affairs in the Fox household first reached a crisis, and certainly from the broad period when Mill and Harriet

---

[14] J. Mill 1820: 12–3; J. S. Mill 1989: 95; J. S. Mill 1961: 184.

[15] Bentham 1843: 349, 352, 544–5; Folbre 1992; Crompton 1985: 38–53; Walker 1840: 177; *Medical Press and Circular*, 1869(1), 367–8; *Public Health*, 1868(1), 191–3.

[16] J. S. Mill 1972: 1693.

[17] J. S. Mill 1984: pp. lviii–lx, 35–49, 375–7.

were in much the same predicament as Fox and Eliza, that is, at the
outset of their affair and the rupture with Harriet's husband. Hence
the title is misleading, in the sense that the authors are not undertak-
ing wedlock, but dissolving it, and they deal with the evils rather
than the rewards of the married state. On Harriet's side the rejection
of marriage is actually so sweeping that it would seem to preclude
even a sexual bond with Mill—confirming what was apparently the
surprising truth about their affair, that it was not consummated.
Harriet's marriage to Taylor may have been, for once, the kind of
experience which modern wisdom holds to be universal in the
period, a matter of her having bartered 'person for bread' in a physically one-sided relationship: 'all the pleasures such as there are being
men's, and all the disagreeables and pains being women's'. 'I have no
doubt', says Harriet, 'that when the whole community is really educated . . . no one would marry.'

This is in some respects in direct disagreement with her partner's
statement. Mill is just as keen on the dissolution of marriage, of
course, but for him this is a transgressive act sanctioned for the 'highest natures' by the imperative of a finer relationship: 'to such, every
tie . . . which restrains them from seeking out and uniting themselves
with some one whom they can perfectly love, is a yoke to which they
cannot be subjected without oppression'. And these 'highest natures
are of course impassioned natures'. But not in the sense one might
suppose—for it emerges that the rejection of marriage is also the
rejection of an institution 'made *by* sensualists, and *for* sensualists',
whose 'purposes never could have entered into the minds of any to
whom nature had given souls capable of the higher degrees of happiness'. And, while on one hand the transgressive unions permitted to
higher natures are thus drained of physicality, on the other the old
sensual, matrimonialist code of the rest of mankind is tacitly allowed
to remain in place, as 'more suitable than at first sight may be supposed by those whose nature is widely different'—for even the 'best
popular morality' knows no higher good, given the low 'degree of
development yet attained by average human nature'.

So Mill's position, while it gives a priority to illicit heterosexual
bonds, comes close to Harriet's after all, through the logic which
equates superiority to codes with superiority to sexual drives.[18] The
equation is reminiscent of Shelley, or at least of one reading of the

---

[18] See also Mendus 1989.

latter's views on sexual ethics. Probably the poet's most famous published remark on these questions was the tag that 'chastity is a monk-
ish and evangelical superstition', which appeared in the 'Even love is
sold' footnote essay in *Queen Mab*. The footnote at large argues for
freely dissoluble unions, and certainly had an inflammatory reputation, so much so that Robert Owen found himself greatly embarrassed when it seemed that socialist propaganda endorsed it. But
Shelley's attack on chastity goes on to say that it 'is a greater foe to
natural temperance even than unintellectual sensuality', and with
that 'natural temperance' we realize that Shelley is far indeed from a
libertarian view. In an obscure essay on Greek ideas of love, not published until 1833, Shelley had in fact changed the emphasis in the
triad chastity-temperance-sensuality by asserting that 'an enlightened
philosophy, while it must condemn the laws by which an indulgence
in the sexual instinct is usually regulated, suggests, however, the propriety of habits of chastity in like manner with those of temperance'.
Philosophy takes this austere view, according to Shelley, because of a
decay of the relative standing of sensual pleasure, compared with
spiritual, as civilization advances: 'the gratification of the senses is no
longer all that is sought in sexual connection. It soon becomes a very
small part of the profound and complicated sentiment which we call
love'.[19]

On the face of it, Shelley offers a most promising route leading
from the progressive to the anti-sensual. Shelleyan radicalism—
unlike the mere high-mindedness, as one may see it, of Hazlitt
opposing Malthusian quietism on sexual drives, or of the anti-
monarchical faction crying out on the morals of the Prince Regent—
is a case where a liberationist sexual code carries within itself an anti-
sensual principle. But I cannot report that the peculiar Shelleyan
formula makes a significant showing in the literature of respectable
radicalism in the Victorian years. Perhaps Shelley's text, already on a
cusp between platonism and sensuality, was too coloured in its
reception by knowledge of the facts of Shelley's transgressive sexual
career.

The intriguing and moving extracts from the diary of Mrs
Robinson which appeared in the legal columns of *The Times* in
1858[20] offer one instance of Shelleyan ideas being accommodated in

[19] Clark 1988: 117, 220–1.
[20] For the divorce proceedings see *The Times* (15 June 1858), 11, (22 June 1858), 11 (27
Nov. 1858), 11, (3 Mar. 1859), 11.

an individual's sexuality in this period, with understandably mixed results. Mrs Robinson was contesting her husband's petition for divorce (one of the first under the new act of 1857) on the grounds of her adultery with the well-known hydropathologist, Edward Lane (he counted Charles Darwin among his patients, and evidence was given on his behalf by his mother-in-law, who was no less a person than Lady Drysdale, unwittingly the mother of the author of the newly published *Elements of Social Science*). The main evidence was a manuscript diary, discovered by Mr Robinson, in which she had recorded her experiences and feelings in relation to Lane over four or five years from 1850 to 1854. Its author had literary ambitions, and the diary is addressed to a 'reader'—and was eventually adjudged by the divorce court to be the fantasy of an unstable middle-aged woman (she was in her early forties), suffering from 'uterine disease' and 'unchaste desires'.

I take the diary to be a true account of events, which makes Mrs Robinson's situation in 1858 a wretched one: for she had been rejected by Lane after a few furtive acts of intercourse, and was then obliged to agree with his denial of the affair (and hence of her own passionate record of it) in defending herself against her husband's petition. She quotes in the diary from the cardinal Shelley poem of triangular love, *Epipsychidion*, and her feelings for Lane were both high-toned and strongly physical. 'Heaven could not be more blessed' than the moments of their lovemaking, but she is glad when he tells her that her attraction to him had never been 'indelicately expressed'. Later, when Lane's coldness is apparent, she records the turmoil of her feelings. Her phrasing may owe a good deal to melodrama, but there is a tension in this list of emotions which is not stagey: 'anger, sensuality, helplessness, hopelessness overpower and rend my soul, and fill me with remorse and foreboding'.

Another writer with progressive credentials in whom sensualism and anti-sensualism are strikingly conjoined, but whose influence on English opinion can be traced with rather more confidence, is Rousseau, especially the Rousseau of *Emile* (1762). *Emile* expounds ideas on education through the words of a tutor-figure and his hypothetical pupil. It is an exasperating blend of pioneering insight and fatuity, with a great deal to say about sex. Well over half of its length is devoted to Emile's pubescence, his and Sophy's mutual sexual awakening, and their courtship. The book culminates with their marriage: this is 'the crown of my labours', says the tutor-figure, while his advice about 'the bridge between the child and the man' he regards as the most valuable he has to impart. Underpinning all this material is an extreme version of the view that libido is modifiable. According to Rousseau, sexual desire in the male before the age of 20, at the earliest, is an artefact of civilization. Even after this point a complete continence is perfectly achievable, because male sexual desire is so much a matter of mind and imagination: 'as I have said again and again, it is the imagination which stirs the senses. Desire is not a physical need; it is not true that it is a need at all.'[21]

Sexual restraint is extolled in the highest terms in *Emile*. The book ends with a positive orgy of abstinence, as Emile is first made to travel for two years before he can marry Sophy (by which time he is 24), and is then denied access to her bed on their wedding-night. This is on the instigation of the tutor, who throughout this strange hybrid of novel and educational tract manipulates and monitors his charges with an alarming thoroughness. But the almost incredible asceticism of the book's doctrines is not straightforwardly anti-sensual, indeed it has a pro-sensual aspect. One of many threads in *Emile* which anticipate English nineteenth-century ideas is the emphasis put on the erotic satisfactions of marital love—enhanced as they are by abstinence before, and restraint during, the ideal marriage (and one aspect of this restraint, to Rousseau's credit, is the right of the wife to decline intercourse unless she feels desire). In all phases of our sexuality, according to Rousseau, our actual *enjoyment* is enhanced by postponement, even by deprivation: in a remarkable paradox, the tutor assures his pupil, who is panting for the consummation of their love after three months' engagement to Sophy and some intense but limited caresses, that he is as good as sexually fulfilled: 'before tasting the pleasures of life you have plumbed the depths of its happiness . . . You have tasted greater joys through hope than you will ever enjoy in reality.'[22]

Rousseau's educational theories had their enthusiastic English proponents in the latter eighteenth century, most notably Thomas Day and, more briefly, his friend Richard Lovell Edgeworth, the father of the novelist. But this kind of systematic propaganda (as opposed, say, to the more casual homage paid by Cobbett in the fifth section of *Advice to Young Men*) seems to have been a transient fashion, in which the sexual aspect of Rousseau's pedagogics did not

---

[21] Rousseau 1974: 298, 378, 439.   [22] Ibid. 411.

figure very largely. One of the few unmistakably Rousseauesque treatments of youthful sexuality is the work of a figure in the Edgeworth circle, the Bristol physician Thomas Beddoes, who married another of the Edgeworth daughters. His *Hygeia* of 1802 accepts Rousseau's premiss that sexual ideas and information can arouse sexual drives before their natural biological onset. His solution is that a child's mentors—either its parents, or its teachers in the 'Healthy School' projected by Beddoes—should pre-empt the wrong kind of sexual learning with sedate and factual instruction in the appropriate anatomy, physiology, and pathology. This might include a visit to a hospital to see patients with venereal disease, a proposal to be found also in *Emile*.[23]

But the *Hygeia* seems to have been one of the least successful of Beddoes's many enterprises, and it is doubtful if its several anticipations of later nineteenth-century English ideas—its concern to thwart youthful masturbation through a kind of moral quarantine, for example, and the theory of accelerated puberty—show that the latter have a direct ancestry in Rousseau. On the contrary, in some minds Rousseau may have been associated with libertarian doctrines because of the great weight of sexual reference in *Emile* (what Burke called its 'pedantic lewdness'), regardless of its tenor. Mary Wollstonecraft, in *The Rights of Women*, consistently holds him up as a base sensualist.

The *Letters on Marriage* of 1812, by Henry Kitchener, is a wide-ranging and interesting text entirely about sex (it also calls itself *Letters on the Sexual Intercourse*), and is studded with quotations from *Emile* and from Beddoes. In his warnings about masturbation Kitchener repeats a specific doctrine from *Emile*, that this practice is more deplorable than 'illicit commerce with women'. He follows Rousseau on the plasticity of the libido (young middle-class men will sometimes imagine that they have sexual needs which must be relieved by a visit to a brothel, when these have really been created by louche behaviour in their own social circle), but he has more respect for sexual gratification than the Frenchman. He takes explicit exception to Sophy's repelling of Emile in the bedroom ('a more obnoxious passage I do not ever recollect to have met with on this subject'), and seeks to exonerate Rousseau by supposing that he means 'a *momentary* kind of toyful repulse'—which he surely did not.[24] But, again,

---

[23] Beddoes 1802: 56.   [24] Kitchener 1812: i. 22, 289, 325.

Kitchener was a minor figure, with no apparent impact on contemporaries: nothing is discoverable about him today.

II

Many Mansions

Rousseau's *Emile* is an endocrine gland in the system of English nineteenth-century anti-sensualism. Its ideas are so commonly echoed (the notion that the first onset of sexual impulses in an individual is culturally determined, in particular, has numerous variants, including William Acton's 'semi-continence') that one can hardly doubt its influence—but direct channels seem to be almost nonexistent. Even if Rousseau's few avowed English followers had no influence and attracted no disciples, their views are of interest, however, as an indication of where minds of this general intellectual persuasion could be led. For there is a coherence to progressive thought, a set of assumptions about human beings and of goals for society, which may be expected to lead to common views on questions of sexual morality. By comparison Christianity, even Protestant English Christianity in the nineteenth century, is incoherent.

This may seem an implausible remark, but where did the common ground lie? Obviously not—if one looks across the range of Christian denominations—in ecclesiastical authority or in liturgy, and not even in a basic metaphysics of supernatural agents, transcendent realms, and immortal principles. Was Jesus divine? Did Hell exist? What was Heaven like? Did the personal soul survive? There was no one 'Christian' answer to these questions. What there was in the way of an agreed Christian narrative (Fall, Redemption of Man by Christ, Salvation for believers) was interpreted in so many ways that it may as well not have been available. The mileage that two authors of the 1870s, C. Maurice Davies and J. Ewing Ritchie, were able to find merely in the description of contemporary sects indicates not only the multiplicity of the religious scene, but the isolation in which some religious groups operated.

Christianity does have its bedrock, which is the Bible. This text certainly contains injunctions of various sorts on sexual morality—and there is also what can be called a Christian moral aura, which derives from the practice of men and women trying to live out biblical teachings, especially in the Dark Ages and the medieval period.

This aura is anti-sensual, and an important presence: it does shift the moral centre of gravity of Christianity in the direction of sexual decorum and restraint, and heterosexual monogamy. But this is only a tendency of thought, and a broad one at that. The Bible itself, if appealed to, allows a range of options within this tendency, and even beyond it.

Masturbation and homosexuality appear to be condemned by the Bible, though chiefly in the Old Testament; the latter implicitly sanctions polygamy, concubinage, and adultery. The New Testament is rather more obviously severe on these topics, so much so, indeed, as perhaps to recommend celibacy as the highest ideal—but this was an unwelcome idea in nineteenth-century England, and St Paul's words about this were usually given a less austere complexion. Christ himself had apparently set a high standard by condemning mere heterosexual desire, if it was illicitly directed, but he had also most memorably forgiven the woman taken in adultery. Or had he? William Paley doubted it. The fact is that biblical texts were commandeered on behalf of many different points of view in nineteenth-century England, without much concern for context or likelihood of interpretation. This is reminiscent of the Malthusian debate, in which the most prolific use of scripture actually emanated from those who wanted all young people to marry and have numerous children ('Anti-Malthusians quoting their Bible against palpable facts are not a pleasant spectacle', grumbled Carlyle).[25] In other words, the beliefs of a sect or religious grouping on sexual questions tend to come first, and the scriptural warrant second.

One might suppose that there was a coherence to sectarian ideology, however, even if Christianity at large was incoherent by comparison with political and social reformism. It is a commonplace, for example, that Evangelicalism—meaning the less intellectual dissenting groups and the low end of the Anglican Church—was more anti-sensual than the broad centre of Anglicanism. But in some important respects this is a misleading description of the Evangelical position. One must notice how the logic of a sectarian ideology can work. A belief in sin and a concern for salvation, for example, do not necessarily entail antagonism towards the sexually transgressive: many Evangelicals found that the logic of their beliefs pulled them in just the opposite direction. One of the most vehement critics of

25 Carlyle 1899: 201.

Malthusian moral restraint was the pseudonymous 'Simplex', who believed as a good Calvinist that humanity was fallen and sinful, but also that it was evil for sexuality to be restrained. It is true that at the denominational level one can speak of structured Christian beliefs in a way that is not possible for the English Christian community as a whole in the nineteenth century (and I shall be taking such an approach in this chapter), but the place of sexuality in these systems of belief was sometimes not what it is usually taken to be.

Sectarian beliefs do say something about communal Christianity, of course, because denominations varied in size, and there were some very big groupings and some very small ones. In this sense the impact of religiously based sexual codes on the society can actually be assessed better than that of secularist ones. Progressive thought may be coherent, but we have little idea how much support it commanded. We can venture to predict what a man or woman of advanced views would have thought about sex, but we do not know how many such individuals there were (they would admittedly have encountered some major bifurcations in ideology, as I will describe, but these were clearcut alternatives). Christianity may be incoherent, but we know roughly how many people assented to its various versions. In what follows I shall first comment on some of the more unexpected lines of thought on sexuality, springing from Christian premises, to be encountered in the period. Then, with lessons learnt from this inquiry, I shall discuss the religious-sexual codes which were dominant in the English nineteenth century, to judge from the facts of religious allegiance.

In Protestant England the centre of gravity of Christian moral thinking was shifted somewhat back in a pro-sensual direction by an almost universal distaste for chastity as a deliberate and dedicated condition (and especially so in a climate in which, as I have argued in the first part of this study, members of both sexes consciously valued heterosexual marriage for its erotic satisfactions). The ideal of religious celibacy, so important for primitive Christianity, and so respected still in the Middle Ages, had only a tenuous footing in the religious life of nineteenth-century England. The clergy of the Roman Catholic Church were, of course, committed to celibacy; around an eighth of British churchgoers in the first half of the century were Catholics, and around a quarter in the second half (representing up to 4 per cent of the whole population)—with a heavy concentration among the Irish population of Liverpool. With the

Protestant majority, clerical and monastic sexual codes (supposedly issuing either in clandestine depravity or in psychological and physical damage) remained an insurmountable objection to Catholicism. Blanco White, an important spokesman on these questions, was able to draw with telling effect on his personal recollections as a young man destined for a clerical career in Spain, writing of two sisters who died young in nunneries, and of the moral destruction of his fellow students ('not one was tainted by the breath of gross vice till the church had doomed them to a life of celibacy').[26]

The urbane, Evangelically minded lay theologian Isaac Taylor mounted attack after attack on the historical stronghold of Christian celibacy, the early church, arguing that natural human 'pudicity' had, very early in the Christian era, been taken to extremes by a cruel fanaticism. The claim was given a topical edge by Taylor in 1839, in his *Ancient Christianity*, as a warning against the intentions of the Oxford Movement. Though clerical celibacy was probably not thought a workable prospect by the leaders of this movement it was certainly among their hopes that monastic orders for men and women would re-establish themselves in Britain. Pusey promoted a nunnery revival as actively as he dared: the first sisterhood was founded in 1845 and about 1,300 women had taken vows by the end of the century. How irrevocable a commitment to celibacy was involved in these vows was a matter of fierce claim and counter-claim. The sponsors of the sisterhood movement denied accusations that they were withdrawing women from marriage and motherhood, but it seems clear that the spirit of their vows, if not the letter, was of a lasting self-dedication to the celibate life.[27]

There is evidently a good deal of moral one-upmanship in the attacks on the Catholic code of chastity. Blanco White is implicitly claiming for Protestantism a higher real standard of sexual conduct than is yielded by Catholic prohibition. Coleridge, who admired White, lamented 'how deep a wound to morals and social purity has that accursed article of the celibacy of the clergy been!' Indeed for one Victorian who became an advocate of free love, but who had been brought up as a typical 'Christian Theist' of his day, the contradiction in the Protestant view was intolerable, and made Catholicism actually the lesser evil: 'I prefer carnal logic to sacred casuistry; and if this denunciation of natural

[26] Blanco White 1825: 130–2, 136. See also Best 1967.

[27] Vance 1985: 36–8; T. T. Carter 1853: 4–5; Goodman 1862: *passim*; Malmgreen 1986: 132–4; Anson 1964: 300–3; Hill 1973: 278–80.

appetites as something bestial be sound doctrine, the plea for irrevocable marriage, as an aid to spiritual purity, is nothing better than a
sham'. According to one (evidently biased) French observer, celibacy
was given such priority as a vice by the Anglican Church that its current theology resembled Ovid's *Ars Amoris*.[28]

Was the logic which led English Protestantism to condemn both
celibacy and extra-marital sex virtually an illogic? In the mainstream
there was no doubt a strong tendency to combine a reflex hostility to
dedicated celibacy with a reflex anti-sensualism, and, as I shall argue
shortly, this was partly because middle-of-the-road Anglicanism was
bound up unthinkingly with another, very ancient creed, of stoical
anti-sensualism. More thoughtful Anglicans may have tried to forge,
and live by, a code that was at once ascetic and hedonistic. Charles
Kingsley is a case in point; despite a courtship that was physically
close and sexually arousing for both partners, he and his bride Fanny
postponed consummating their marriage for about a month.[29]

The rationale offered by Kingsley for this delay is interesting; it
was supposed to stand to full intercourse as the latter stood to the
couple's marriage in the afterlife: 'when we lie naked in each other's
arms, clasped together toying with each other's limbs, buried in each
other's bodies . . . shall we not feel then, even then, that . . . those
thrilling writhings are but dim shadows of a union which shall be
perfect'. There is one religious group whose sexual creed comes irre-
sistibly to mind here, namely the Swedenborgians. As it happens,
there was a small but visible Swedenborgian fraction within the
Church of England. That its members were apparently comfortable,
as Anglicans, with their master's distinctive teaching on sexuality
indicates that English Protestant dislike for celibacy readily took an
actively pro-sensual form.

Marriage in Swedenborg's system is given a considerable emphasis, both as a theological symbol (chiefly via the imagery of
Revelations) and as a literal institution. And not only is human
sexuality thus, in a general way, to the fore in Swedenborgianism, but
the physical content of marriage, Kingsley's 'thrilling writhings', is
quite surprisingly stressed in various respects. The relevant teachings are set down in *The Delights of Wisdom Concerning Conjugial
Love* (the curious adjective has stuck because the 1794 translation,
which is so entitled, was for a long time the only English version).

[28] Coleridge's *Table Talk*, 18 Apr. 1833; Coupland 1878: 13–14; Rubichon 1819: ii. 277.
[29] Chitty 1974: 74–88.

In the Swedenborgian afterlife the human individual is completely the person he or she was in life, gender included. There are marriages in heaven, either freshly made or, for couples who have achieved a fine partnership in life, continuations in a more rhapsodic mode. In these marriages a kind of infertile intercourse occurs, by means of an additional sixth sense: this is 'chaste' and 'internal' but intensely enjoyable. (For Swedenborg chastity is an aspect of sexuality—virgins are not chaste—while this same internal, chaste communion is said to be achieved also in the physical congress of the best earthly marriages: one female disciple insisted that it was quite wrong to 'vaunt the purity of virginity over marriage').[30] The partners in the afterlife are 'naked' and 'in bed they lie copulated'. For Swedenborgians it was a telling consideration that reunion with the loved one would be less than heavenly without sex ('what would become of such a prospect if the husband ceased to expect to meet and unite with the *peculiar feminine virtues* of his wife . . . as . . . beautifully exhibited in the *peculiar feminine charms* of her *countenance* and *form*'). They also candidly counted the pleasures of earthly marriage among the pleasures any self-respecting heaven should offer: 'many as well as myself would not esteem it half a heaven without'.[31]

It seems fair to suppose that such ideas flowed from, or flowed back on, the marital experiences and attitudes of Swedenborgians. At the least, followers were being given an implicit assurance that sexual enjoyment in marriage was as valuable as anyone could ever feel it to be. In the Swedenborgian system of links between the natural and the transcendent realms the joys of earthly marriage were actually a token of heavenly realities: the master taught that 'unless there were ultimate delights, there would be no delights of conjugal love'. As I have mentioned in the first part of the present study, there was a Swedenborgian version of the medically derived theory that women were more thoroughly and comprehensively erotic than men. A woman, says Swedenborg, is 'continually thinking about the man's inclination to her, with the purpose of joining him to herself; it is otherwise with the man' (his views on the sexes are grossly inegalitarian). Souls, in fact, are not merely gendered, they cause gender: 'the distinction of sex originates in the soul'.[32]

[30] Swedenborg 1794: 51, 146–7; Strutt 1857: 62.

[31] *Magazine of Knowledge*, 2 (1791), 193–6; Missionary and Tract Society 1844/No. 6: 19; Spence 1792: 35.

[32] Swedenborg 1794: 52, 168; Hindmarsh 1822a: 263.

It is true that there is plenty of rebuke in *Conjugial Love* for those married individuals who treat sex animalistically, whether in adultery or in marital intercourse which is not infused with spiritual feeling. But Swedenborg is interestingly self-consistent on premarital intercourse: the more it approximates to marriage the more it may be condoned. Keeping a mistress as an expedient to avoid the ills arising from 'too great a check of the love of sex', especially if marriage is still a man's goal, is preferable to casual intercourse. And if marriage breaks down through the illness, depravity, and so forth of the wife (comparable shortcomings in the husband are not discussed) it is 'not illicit' for her partner to take a mistress.[33]

The original *Conjugial Love* dates from 1768, but its lenient teachings on sexuality did not necessarily deter nineteenth-century Swedenborgians. One of the great founders of the movement in England, Robert Hindmarsh, endorsed the idea that there were degrees of viciousness in non-marital sex, and wanted to see prostitution legalized. And it appears that in the early days of the sect's formation in this country one coterie at least, in the Great Eastcheap temple, sought to advocate a system of concubinage such as had been proposed for Swedenborgian communities in America, and to establish the principle that members could seek true marriage outside the legal form, setting aside their wives in favour of a mistress. This group, which included Hindmarsh, was expelled from Great Eastcheap by the other membership. Hindmarsh quite soon returned as the minister of the temple (perhaps because, whatever his theoretical beliefs, he had 'kept aloof . . . from all unchaste practices'), but the rest tried to refound the English Swedenborgian church with their own short-lived venture, the London Universal Society for the Promotion of the New Church.[34]

The New Church was as fissile as most young sects, but it was surprisingly cohesive on sexual questions, given that the often striking doctrines of *Conjugial Love* were a regular part of the ammunition of its enemies—and particularly cohesive by comparison with Owen's and Bradlaugh's followings in a similar predicament, as later chapters will illustrate. (It is interesting that the pamphlet and lecture war on Swedenborgian sexual codes reached its peak at much

---

[33] Swedenborg 1794: 411–22, 432.
[34] Hindmarsh 1822a: 31, 183–4, 194; Lineham 1978: 248–51; M. Paley (1979); *New Church Repository*, 6 (1853), 145.

the same time as the equivalent Owenite controversy, in the late
1830s and early 1840s, though there is sporadic agitation against
them down to 1860s, with John Brindley playing much the same
doughty and long-standing role as he played against secularist sexual
policies.[35] There was no repetition of the Great Eastcheap schism,
despite there still being glimpses in the Swedenborgian press in 1800
of 'the pretence of some, even a society, and also individuals, that the
real conjugal partner is to be sought, and may be discerned here on
earth, whether among married or single'.

In 1818 the London Swedenborgians were plunged in dismay by
the terrible murder of a female member of the Lisle Street temple,
the 18-year-old Mary Minting. The murderer, William Hetch, having
failed to seduce Mary, entered into a bigamous marriage with her. It
is not clear if he was encouraged in this transgressive behaviour by
his own Swedenborgian creed, or by what he took to be his victim's
beliefs. She certainly seems to have found all of it grossly inconsistent
with New Church teaching; she rejected Hetch's advances, left him
when the bigamy was revealed, and he then killed her. The most
intriguing aspect of the story, however, is that *both* parties hoped to
be reunited after death. In fact Mary's trust in this prospect was
believed to have encouraged Hetch to take her life. He welcomed his
execution: 'I am anxious to die and be with that dear object of my
soul; for I am still persuaded that I shall have that unbounded pleasure of enjoying her company in a far superior way'. In the same year
it was recognized by a leading New Church minister that 'worthy and
judicious members', whose wives are also Swedenborgians, nevertheless 'have indulged and *cherished the idea*, yea, delighted themselves
with it, that another person, then existing, and with whom they have
acquaintance, is their destined conjugal partner'. This rather
Wuthering Heights-like way of picturing illicit relationships, taken to
a terrible extreme in the Hetch case, suggests one route by which the
eroticism of Swedenborgian theology could have remained a force
while falling short of the practical libertarianism of the Great
Eastcheap coterie in its effects—its transgressive potential being
subdued by the strange promise of adultery consummated in heaven
(the doctrine of marriages in the afterlife is confidently asserted

[35] Bayley 1838*a*, 1838*b*; Goyder 1839; Pike 1844; 'Examiner' 1839; Noble 1838; Skidmore
1844; *O Bit ov o Chat* 1844; Woodman 1858; Brindley and Woodman 1862; *Intellectual
Repository*, 5 (1839), 293–302, 308–12, NS 1 (1840), 129–30, NS 2 (1841), 494–5, NS 3 (1842),
1–9, 36.

by Swedenborgians of all shades of opinion right through the century).[36]

There *were* overt contradictions over sexual policy (prostitution, condoned by Hindmarsh, was elsewhere denounced), and some of the Swedenborgian literature on courtship and marriage reads as if *Conjugial Love* did not exist. The sect's matrimonialism could lead believers down conservative paths in Victoria's reign: into opposition to birth control, for example, and to divorce (one of the most active opponents of divorce in the Commons, Richard Malins, was a Swedenborgian). However it is notable that a very famous sponsor of contraception, Henry Allbutt, was for a long time a member of the sect: divided from them in due course on birth control, he tells us, but at one on the desirability of early marriage. Other Swedenborgians remained loyal, though greatly irked by 'the smugness and strait lacedness' of many of their co-religionists: still looking to the master's teaching as 'the way to a greater liberty of thought and knowledge on sexual subjects'.[37]

'Liberty' here is not sexual licence, it must be said—rather an escape from both 'hypocrisy' and 'animality', such that 'things before invisible and unutterable will be seen to be as clean as the flowers of the field or as the pleasant butterflies'. The official line against the enemies of the New Church, when they quoted passages from *Conjugial Love* on fornication and concubinage, probably commanded the loyalty even of this disgruntled believer. It was that Swedenborg had been referring to the needs and behaviour of the unregenerate, 'natural' individual: he had analysed sexual licence as a doctor might investigate disease—but had by no means condoned it for the 'spiritual' and 'religious'. 'Where is there a real New Churchman who lives in concubinage?', asked the main defender in the great Accrington pamphlet battle of 1838–44, with a confident ring. But this did leave Swedenborgians free to argue that practices rejected by a repressive society were acceptable, because most men and women were actually part of the unregenerate rump: divorce, state-regulated prostitution (if prostitution there must be), and even

[36] *Aurora*, 2 (1800); Goyder 1857: 95–100; Odhner 1904: 268; *Examiner* (1 Mar. 1818), 138; Churchill 1818: *passim*; Proud 1818: 80; Holcombe 1905: 179; Davies 1874: 194; Hyde 1875: 259.

[37] Missionary and Tract Society 1844/No. 20: 25; Hyde 1875: 256–7; *Malthusian*, 3 (Apr. 1879), 23; Wilkinson 1911: 203–4; Holcombe 1905: 179.

concubinage in countries which forbade marriage by the rites of religious minorities.[38]

There survives a brief personal account of the impact of *Conjugial
Love* on a young Swedenborgian which is illuminating on these questions.[39] The author recalls, for the benefit of young believers who
may be troubled by 'carnal desires and doubts similar to mine', how
in his youth his 'strong passions towards the female sex' had only
been kept in check by his religious convictions, and how this
restraint had been seriously threatened by his first encounter with
the 'scortatory love' section of *Conjugial Love*. 'The strong passions of
my sensual nature were again brought into powerful excitement' by
these pages, apparently so inconsistent with the rest of the text, he
tells us. He did not fall, however, and came eventually to see that
Swedenborg was only issuing a permission to the 'natural' man.

The experience of being tempted by Swedenborg's text he presents interestingly as a matter of his own mind working hard to
square his desire for casual sex with the ambiguous encouragements
which the master seemed to be offering: 'I tried it years before I was
married; I endeavoured and *earnestly desired* to see how it ['fornication'] might be allowable to me; and this not once only, but several
times . . . Evil spirits presented to view the *possibility* of enjoying the
delights peculiar to the lowest degree [of 'natural man']'. The notion
of 'evil spirits' partly explains why there is an odd feeling of distance
from personal impulse here: this is a characteristic effect in
Nonconformist spiritual autobiography. But even if one interprets
the endeavour 'to see how it might be allowable to me' as sheer
youthful lust, it still appears that this young believer needed the complete sanction of Swedenborg's text before he would surrender—
although his lust had been largely provoked by that text in the first
place. It is a testimony to the power of Swedenborgian discipline, but
also to the strength of a young church member's conviction that his
religion could not, in the last resort, be permitting what it appeared
to be permitting in the way of sexual freedom.

Instructive too is the way the author recalls having counterbalanced
his young urges. For this purpose he used to dwell in his mind on the
delights of marriage, resorting frequently to a biblical text which he
recommends to his readers. This text (Proverbs 5: 15–19) might seem

---

[38] *Intellectual Repository*, 5 (1839), 308–12, NS 2 (1841), 493–5; Bayley 1838*a*: 8–9; 1838*b*:
25; Woodman 1865: 190–217, 241, 293.
[39] *Intellectual Repository*, 5 (1839), 293–302.

about as unlikely to 'quell' a sexually impatient young man's 'loose desires' as any that could be found, but this is what he tells us it often achieved. Even if the whole reminiscence is fabricated, in fact, it is an index of how authentic was the conjunction, for this kind of temperament, between matrimonial eroticism and extra-matrimonial chastity that the author can prescribe the following as a textual cold shower: 'Let thy fountain be blessed: and rejoice with the wife of thy youth. | Let her be as the loving hind and pleasant roe; let her breasts satisfy thee at all times; and be thou ravished always with her love'.

The moral psychology may seem paradoxical, but this is surely not the 'sacred casuistry' on married sex which we have heard orthodox Christian belief being accused of. Swedenborgians liked to think, in fact, that their position was more whole-heartedly anti-celibate than the Anglican one. They argued that Anglican liturgy and teaching failed to put the positive case for marriage (as opposed to simply denouncing extra-marital sex), and that in Protestantism generally, despite the historical lessons conveyed by Isaac Taylor, it was still 'almost universally received . . . that marriage is an institution suited only to our grosser nature'.[40] On the other hand, this scarcely rings true for the general mentality of nineteenth-century England, and it is not surprising that Swedenborgianism, with its erotic matrimonialism, had an ideological niche within the Church of England. The sect itself was tiny if buoyant (the New Church doubled its membership, between 1825 and 1850, from about 1,300 to about 2,600 individuals, and went on growing thereafter),[41] and its significance is considerably enlarged by these signs of affinity with Anglicanism.

To start with, one segment of English Swedenborgianism originated within the Church of England and never left it. John Clowes, incumbent at St John's in Manchester's Deansgate, was a pioneer of equal stature with Robert Hindmarsh, and the man chiefly responsible for the circulation of Swedenborg's text and teaching in English. He was the translator of Conjugial Love, and his own inserted 'Apology' for the 'scortatory love' chapters actually does nothing to remove the impression that Swedenborg is tolerant of extra-marital sex when it assists true matrimony. By 1783 there was alarm over Clowes's unorthodox beliefs and Bishop Porteous (then Bishop of Chester, later the fighting Evangelical Bishop of London) was appealed to by local Anglicans. Porteous remonstrated with Clowes

---

[40] *Intellectual Repository*, NS 3 (1842), 2–4; Bruce 1850: 9; Woodman 1865: 137–43.
[41] Lineham 1978: app. 1; Ritchie 1870: 252.

and probably extracted some kind of assurance that he would promote Swedenborgian doctrines less in the future.

If Clowes obeyed this injunction in his pulpit he seems to have
thought he had permission to teach Swedenborgianism in other ways
(it was later said that Porteous had been 'by no means dissatisfied or
displeased with his opinions'), and he continued with his writing,
publishing, and itinerant preaching in the Lancashire district. The St
John's faction remained a vigorous non-separatist Swedenborgian
group within the Church of England until 1861. Clowes himself was
offered a bishopric in 1804.[42]

Sexual doctrines are not known to have figured in the events of 1783
(it was Clowes's anti-Trinitarian teaching which gave the chief
offence), and Clowes was still some years away from the translation of
Conjugial Love. However, the latter first started to appear in serialized
form in the *New Jerusalem Magazine* in 1790, and this was the organ of
the breakaway group in the Great Eastcheap temple of advanced
views. Clowes also published in 1814 a pamphlet of matrimonial
advice, *The Golden Wedding Ring*, which was reprinted right through
the century. It is a confident Swedenborgian celebration of marriage as
a spiritual and physical relationship. There is the characteristic logic
whereby the pleasures of earthly marriage are both a token of the
human capacity for bliss and an assurance that heaven will be the
occasion for erotic joys transcending all our previous experience:

all perceive, and allow, that this love, regarded merely as a natural love, is
the most delightful and gratifying of all loves, having more natural bliss
annexed to it than any other love. All, therefore, can perceive . . . that when
this love is . . . directed upwards to its Divine Source, and enjoyed under the
influence of that Source, it then becomes admissive of spiritual and heavenly
blessedness, as far superior to those which other loves enjoy, as its natural
delights are superior to the natural delights attending any other love.[43]

Clowes's congregation at St John's had a much higher proportion
of middle- and upper-class members than the Swedenborgian movement proper in the area, which Clowes had also helped to bring into
life. In the rest of Lancashire the New Church was more of a working-class sect even than the Methodism of the region—with over 70
per cent of its membership coming from these levels. The London
branches of the New Church, by contrast, looked much more like St
John's in social terms: less than a quarter of their members were

[42] Lineham 1978: 94–105; Ritchie 1870: 257.   [43] Clowes 1888: 17.

working-class, and over a third were in the trading professions. English Swedenborgianism in fact had two main geographical focuses—semi-industrial Lancashire and Yorkshire, and London— and two kinds of social profile, with probably two main kinds of subsidiary religious affiliation: Anglicanism and Methodism. It seems that about a third of all Swedenborgian converts had an Anglican background, and much the same proportion a Methodist one.[44]

Methodists were perhaps chiefly drawn to the New Church by its mysticism. This was an aspect of religious experience not at all catered for by Wesleyan Christianity, even though one of the great inspirers of Methodist conviction, William Law, had written in both an evangelical and a mystical vein. An attraction of Swedenborgianism for Anglicans may have been the pro-sensualism which flowed generally from the stress on the human member of the Trinity and from the corporeal, physical character of Swedenborg's heaven, but also from the teachings on conjugal love. We have noticed the affinity between Kingsley and the Swedenborgian temper in the sexual domain. Swedenborgian doctrines about love and the afterlife may have distilled feelings which were more generally and hazily abroad, affecting people of all kinds of religious persuasion. Laurence Housman's widowed grandmother once asked her daughter: 'do you think . . . that when I'm with my old man again, we shall know each other as we used to do?' Coventry Patmore's poem 'The Angel in the House', which is often cited as a normative statement of Victorian views on the relations of the sexes, was written under the direct influence of *Conjugial Love*, which Patmore regarded as 'the finest work ever written' on the subject. Elizabeth Barrett Browning was another enthusiast for Swedenborg's matrimonial doctrine. I have mentioned the echo in *Wuthering Heights* of the idea of a deep, transgressive sexual affinity which will be consummated in the afterlife (Patmore expected to meet his first wife in heaven, although he married twice after her death). When Keats writes in a letter of November 1817 of seeing a beautiful face and knowing that we shall see it in the 'hereafter', only more beautiful, he sounds like a member of the New Church. Coleridge's earnest dictum, in *Biographia Literaria*, that 'there is a sex in our souls' was often quoted by Swedenborgians.[45]

[44] Williams-Hogan 1985: 682–92; Lineham 1978: 125.
[45] Housman 1937: 52; Patmore 1858; Reid 1957: 75–81; Trobridge 1906; Coleridge 1983: ii. 174.

In the 1870s, allegedly, 'many clergymen' in the Church of England were tacit Swedenborgians. There is one detectable case of an Anglican minister with no formal Swedenborgian connections but thoroughly Swedenborgian ideas on sex, the Reverend G. D. Haughton. His *On Sex in the World to Come* (1841) is devoted entirely to arguing for the survival of gender and for the prospect of 'all the fervours of a private attachment' in heaven. The power of the sexual drive, in both men and women, is the warrant that it will find a more satisfactory outlet in heaven than it generally does on earth. The idea that we will lapse into an 'unimpassioned universal benevolence' after death is a wretchedly 'emasculated' one.[46] Haughton's book was reviewed in the *Monthly Review* with only a slight sense of alarm at its approach to the 'sensual' and the 'offensive'. He himself seems to have felt that he was situated quite centrally in ideological terms. He has warm words for Coleridge as well as Isaac Taylor, and for Edward Irving as well as the Oxford Movement. This indicates what a broad swathe of nineteenth-century religious opinion could be enlisted, at least by a sufficiently single-minded writer, in support of a pro-sensual position.

The main omissions in Haughton's cast of supporting opinions are perhaps Methodism (from which some of the fiercest attacks on the New Church emanated, and which I shall discuss later in this chapter), and Unitarianism. Swedenborgian Christianity is the antithesis of Unitarianism, in the sense that it insists on the divinity of Christ as much as the latter insists on his humanity. This confers on Christ a dual nature which Swedenborgians found very congenial to their brand of physical transcendentalism, and it is precisely in sex, where the gap between the physical and the spiritual seems to be at its widest, that this duality is best registered. 'The reason why in the Word such frequent mention is made of the intercourse between the sexes, and of those parts of the human body, which are more particularly dedicated to that office', argued Hindmarsh, is that in Christ 'every portion of his body, even to the very sensual and corporeal principles, was at length completely identified with the divine nature'.[47]

When Joseph Priestley sets down his objections to the New Church he dwells only briefly and primly on its sexual doctrines. Marriage is ordained for the propagation of the species, and so, asks

[46] Davies 1873: 158; Haughton 1841: 33, 114–16, 185.
[47] Hindmarsh 1822a: 196–7.

Priestley, what purpose can it have in the afterlife?—an attitude in very stark contrast to the candid hedonism of Swedenborgian teaching. Priestley sees the point that in a worthwhile heaven of his sort we must also lose the important marks of sexuality, in a kind of metaphysical version of Godwin's utopia of reduced drives (the 'human frame will, no doubt, be considerably altered'). This aroused indignation: 'my good Sir, do you think that after death men and women will . . . be made hermaphrodites? . . . Men are men, and women are women; and until God himself tells me they shall be hermaphrodites, I will not believe it—and that, I am sure, will never be'.[48]

Unitarianism as such could perhaps only lead to pro-sensual attitudes via its tendency to reformism, the route William Johnson Fox wished it to go down. Nevertheless, a famous Unitarian seems to have given an involuntary impetus to an outcropping of pro-sensual Christian ideas in another sect, to judge from Edward Irving's remarks on the debt he owed to Coleridge.[49] Irving, a Scottish Presbyterian minister who had an intense, brief fashionability as a London preacher in the 1820s, represents another case of sectarian belief achieving an unexpected resonance outside sectarian confines, and outside the usual social range of Nonconformist enthusiasm. Irving was to move further into religious idiosyncrasy than any Swedenborgian, for he seems to have been the first Christian leader in modern times with 'pentecostal' doctrines, and followers who were observed 'speaking with tongues' under the supposed influence of the Holy Ghost.

But in the period of his great popularity with middle-class, metropolitan audiences Irving's distinctive teaching was of the sinfulness of Christ—a theological position he could well have been brought to by Unitarian influence (although Coleridge, by the time of his association with Irving in the early 1820s, was theoretically back in the Anglican fold). Irving taught that Christ, in taking on flesh in his birth to Mary, had also taken on the sinfulness inherent in all post-Adamic flesh: he only became sinless through his baptism and reception of the Holy Spirit. Thus Irving's treatment of the gospel story gave it a close bearing on the situation of his audience, whose members he was seeking to awaken to a new spirituality. The result might have been a Calvinist antinomianism, that is, an assurance to his followers that their salvation was not in doubt, and could not be

[48] Priestley 1791b: 43; Proud 1791: 57–9.   [49] Dollimore 1983: 46.

prevented by any degree of sin, once they had experienced conversion. But Irving's thought tended increasingly to a different conclusion: not that individual faith was supremely efficacious, but that human depravity had been completely atoned for by Christ.

This was Scottish Calvinism taking a strange turn (though one first made by Irving's Scottish mentors)[50] in the direction of a charismatic and adventist religion in which election, damnation, and salvation had little role. Irvingite Christianity offered a Messiah who was not sinless, and a humanity which was not depraved. In 1832, when W. J. Fox was separating from his wife and sowing the seeds of disruption in his congregation at South Place, Irving and his followers were evicted from their Regent Square chapel a few hundred yards to the north-east: their formal heresy was the doctrine of Christ's sinfulness. They founded a wealthy and vigorous denomination, the Catholic Apostolic Church, though Irving himself died in 1834. His career and theological novelties were by now being reproduced in some measure by a man who took pro-sensual revivalism further, and into the Victorian era: Henry James Prince.

Prince at first set out on a track rather different to Irving's. He came from a well-to-do background in Bath and trained for medicine; his initial period of religious vocation was as an Anglican ordinand and minister. But in the 1840s his spectacular impact on congregations in rural Somerset and then rural Essex, plus the doctrinal content of his teaching, led to episcopal reprimands and prohibition. Prince continued to exert his powers as a preacher and prayer-leader with more socially elevated independent congregations in centres such as Brighton and Weymouth, but the real tendency of his work was shown with the founding of a small religious house, called the Agapemone, in the latter resort. By the early 1850s, with large donations from wealthy followers, the Agapemone was established in substantial premises—some purpose-built, some existing farms and cottages—on a large estate at Spaxton, near Prince's original Somerset parish of Charlinch.

The census return for 1851 showed sixty-five persons resident at the Agapemone, of whom twenty-eight were servants (some of these may have also have been believers). The figures were much the same in 1868, though it is probable that the number of inmates rose and then fell in the interval. By the latter date there were also, it was said,

---

[50] Dollimore 1983: 87; J. F. C. Harrison 1969: 97.

some 600 individuals of Princeite beliefs, many of them clergymen, elsewhere in England and Wales. Three Princeites in the ministry of the Welsh church are known for certain, one of whom became a canon at Llandaff. The movement was vigorous enough to be able to build an impressive church in London, at Clapton, a few years before Prince's death in 1899. Here the installation of Prince's successor, Hugh Smyth-Pigott, in 1902 attracted a gathering of some 6,000 people, though many of these were certainly just curious or hostile. By the late 1860s maintaining contact with the discipleship at large, though the latter may originally have been built up by Prince on itinerant preaching trips, had become the main practical work of the Agapemone. This was an adventist sect, awaiting the consummation of a judgement which was already under way and in which the Princeites had already been set aside for salvation.[51]

The passivity of the sect—no conferences, no chain of temples, no periodicals (the *Daily Telegraph* at the time of Prince's death called it 'the community that has no history')—does make obscure not only the question of its scale, but also the general impact of the Princeite movement. On one hand Prince's teaching appears strongly anti-sensual, with abstention from sexual intercourse as a cardinal doctrine. It is well attested that those within the Agapemone not only could not marry, but were required to live separately from their spouses if already married (a rule which aroused some disgust outside as 'Malthusian'). There was a pattern among the male leadership of the movement, Prince included, of marrying older women in supposed obedience to a higher call, and when one of these wives became pregnant she was expelled. The 'truly called' outside the community, equally, are said to have lived 'as virgins'.[52] Local information in the village of Spaxton was that during the 1860s no children were born to the many couples living in the Agapemone. (It is tidiest to assume that this belief, if accurately reported in *The Times*, did not apply to Prince himself, since it is almost certain, as we shall see, that he fathered a child on a disciple in 1856 and that this was well known outside.)

But for local minds the significance of the Agapemone's lack of children could have been rather different: that the Agapemoneites

---

[51] *The Times* (26 May 1851), 8; Dixon 1868: ii. 248, 330; O. W. Jones 1970; Mander 1976: 137; *Illustrated London News* (29 Mar. 1851), 253.

[52] Dixon 1868: ii. 249, 311, 330; *Chambers Encyclopaedia*, i (London, 1860), 72; *Illustrated London News* (28 July 1860), 81; Montgomery 1962: 184.

were ostensibly chaste but not literally so, and had foully concealed the results of sexual intercourse both with their spouses and with others. There were stories abroad of 'dead babies buried in the garden'. In other words, this apparently anti-sensual sect may have been rather thoroughly and resolutely pro-sensual. The rumour that the Agapemone was the site of free-love practices (for which its name was, after all, an excellent warrant) had a wide currency from early on. A French Saint-Simonian missionary and a militant French Catholic were, respectively, attracted and disgusted by this supposed aspect of its operations. 'Mormonism' and 'free lovers' were terms confidently bandied in the press. 'Agapemone' was co-opted by the *Anonyma* author in 1864 to describe certain meeting-grounds of the *demi-monde*, and in that year the term was also sarcastically applied to particularly crowded sleeping accommodation in rural Bedfordshire.[53]

What was the real nature of the relationship between austerity and licence in the Princeite movement? All modern discussions of the Agapemone as an institution have been in a popular vein (though there has been some serious comment on the theology of the sect).[54] These have dealt with its moral contradictions crudely, by isolating Prince as the libertine behind the walls. In this picture the celibacy of the other inmates, with each other, is a fact, but they are also agents of Prince's lust, mesmerized into acting as odalisques or eunuchs in his harem, depending on their sex: in Aldous Huxley's phrase, he was a 'snake-charmer'.[55]

The Times does speak in 1860 of one of the leaders of the sect being 'deprived of his wife for six weeks or thereabouts' because of 'his lord's assumptions', but the later currency of the harem myth is surely due to William Hepworth Dixon's account of the Agapemone in 1868. In the small array of first-hand information about the sect this is by far the fullest source, as well as being itself a vivid and accomplished price of popular writing, and it has dictated the pattern of all later treatments. Dixon, a journalist and editor of the *Athenaeum*, visited the Agapemone in 1867, and then devoted several chapters of *Spiritual Wives*—his book about modern religious

---

[53] *Taunton Courier and Western Advertiser* (27 Aug. 1856), 9; *Bridgwater Times* (2 Oct. 1864), 3; McCormick 1962: 40; Kervigan 1860: ii. 146; *The Times* (29 Sept. 1856), 12, (18 May 1860), 5, (15 Aug. 1860), 5; *Anonyma* 1864: 79; Hunter 1865: 146.

[54] e.g. Ronald Matthews 1936: 170–8.

[55] Huxley 1936: 162.

polygamy—to a fragmentary, atmospheric, and tantalizing, not to say titillating, description of the place and its occupants. He was above all fascinated by the legend that in late 1855 or early 1856 Prince had performed a ritual copulation or pre-copulatory ceremony with a young inmate, Miss Patterson, in the community's chapel, in front of other disciples.

Dixon was not the first man to make public some such story. Prince's liaison with Miss Patterson probably led to the birth of a child in the autumn of 1856, and at this point events became a matter of such scandal that public meetings were held in Bridgwater and Taunton, and even in London, at which Prince's chief disciples gave an account of the sect's doctrines. They admitted candidly that the leader 'took the flesh of woman, and made himself one with it, and all the Agapemone were present at the time'.[56] Ambiguous though this wording is, it is perhaps unlikely that even the most devout of Princeites would have referred so boldly to an act of intercourse performed in public and witnessed by them. In January 1857, in one of a series of booklets which were part of the same exercise of self-explanation by the Agapemone, Prince himself made a remarkable statement about the episode. This clearly suggests that concubinage with Miss Patterson, for some period, followed the ceremony—and thus, perhaps, that what had occurred in the Agapemone's chapel was a transgressive wedding rather than a ritual copulation: 'the Holy Ghost . . . took flesh—a woman—in their presence, and told them that it was His intention to keep this flesh with Him continually by day and by night, and to make it one with Him, even as a man is one flesh with his wife'.[57] Dixon is evasive on what actually took place, and later writers have made up their minds in accordance with how lurid is their vision of the Victorians, or how prurient an experience they are seeking to provide for their readers.

In the following year, 1858, an anonymous pamphlet accused Prince of adultery (his second wife was alive at this date), and specifically named Miss Patterson as his 'violently' taken prey (with whom he was still 'boastingly' cohabiting). Nevertheless, it would seem that the episode did not fix itself in the public consciousness as much as might be expected. When Dixon takes it up ten years later he writes as if he is privy to special information in knowing such details as Miss Patterson's name. His book became a best-seller, and the respectable

---

[56] *Bridgwater Times* (4 Sept. 1856), 3.   [57] Prince 1856–9: 'Voice the Fourth', p. 30.

reviewers who disapproved of its popularity had an interest in claiming that they knew nothing about the Agapemone's practices until Dixon had raked them up. However, it does appear that the general accusation of sexual licence against the Princeites was not a surprise, while the Patterson story was.[58]

Other aspects of the Agapemone legend loomed significantly larger than they would today. First of all, there seems to have been some indignation—perhaps based in traditional anti-monasticism—about the feature of the sect's sexual code that was in a sense antithetical to its alleged licence, namely 'the principle held by the Princeites as to separating man and wife'. Also, on a quite different note, there was considerable sensitivity and indignation about the extent of the donations to the sect ('a party of communists'), especially those from wealthy women who had surrendered their estates to the community when entering it. A highly publicized court-case, which did stick in the public mind for many years, had concerned the abduction of one of these donors by her relations.

The 1858 pamphleteer claimed to be writing to warn potential converts, 'especially of property', and he gives more space to attacks on Prince's heresies than to exposure of his sexual depravity. It is evident that the theology of the sect was also a burning issue for contemporaries, to a degree which may nowadays be hard to understand. (I shall give some account of Princeite theology shortly.) The furore over the Patterson child seems, in fact, to have had a theological dimension; a 'very respectable man' was apparently overheard in 1856 giving this wildly misleading account of the first Bridgwater meeting: "they had a child at the meeting, which was held up, and any one who put his arms around that child had his sins forgiven'. Prince's overt concubinage was denounced because 'the lust of the flesh and the gratification thereof is publicly made a religious ordinance', but equally because 'the Holy Spirit . . . is made to appear as if incarnated in the Head of this sect of Delusionists'. At the Taunton meeting what seems to have provoked real uproar, bringing matters to an abrupt conclusion, was the Princeite doctrine of the judgement:

eleven years ago the Holy Ghost came to this earth . . . He sought not for all men, but a few men, to exhort . . . to prepare and make ready for the coming of the Lord. For the most part the warning was disregarded . . . but the spirit

---

[58] *Mr Prince* 1858: 10; *St Paul's*, 2 (1868), 66–76; *Frasers*, 77 (1868), 655–65.

affected that for which it was sent . . . and the Holy Ghost pronounced that the door was shut.

The message of the Agapemone for the citizens of Taunton was that they were damned, and it was too late to do anything about it.[59]

So however attenuated the popular belief in the Messiah and the Last Judgement at this date it was strong enough to make Prince's messianic claims, and the Princeite talk of an exclusive salvation, intensely disliked. In the late twentieth century we have put the age-old hostility to sects on yet another footing: we perceive them as the cynical creations of cruel and greedy charlatans. This is quite inadequate as an account of, say, the sect led by John Jones which perished so tragically in Guyana in 1978, and probably a serious simplification of the relations between David Koresh and the Branch Davidians of Waco, Texas, but it seems to be the only rationalization which modern society can find for its fear and distrust of such movements.

Dixon's account of his visit to the Agapemone, with its stress on the Patterson episode, has fitted well with contemporary prejudice, and modern popular treatments of the Princeites have represented their leader as an immensely selfish fraud. We apparently have no time for the thought that 'the man who can inspire fidelity such as this is no mere vulgar charlatan', as a neutral assessment, based on some acquaintance with Prince, concluded late in his lifetime. Actually, *Spiritual Wives* should be treated with caution, even at the factual level, for it omits some possibly important details about the Agapemone's history (Dixon later admitted that his research had been fairly cursory). A second Prince, Phillip Alexander Parker Prince—not known to be a relation of Henry James—seems also to have been involved in its foundation: he died in 1862. And around 1851 there had been an earlier 'spiritual manifestation' involving a woman at the Agapemone, which may or may not have been sexual in nature. There are indications that the whole period from 1851 to 1856 (rather than Dixon's rather short chronology) saw an episode of sexual experimentation in the community, and that the public statements of 1856 were by way of marking the sect's renunciation of these events.[60]

[59] *Taunton Courier and Western Advertiser* (10 Sept. 1856), 9, (8 Oct. 1956), 7; *Bridgwater Times* (21 Aug. 1856), 2, (4 Sept. 1856), 3, (9 Oct. 1856), 3.

[60] Sellers 1891; *The Times* (23 Apr. 1862), 6, (29 Sept. 1856), 12, (27 Nov. 1872), 11; O. W. Jones 1970.

Even as it stands, Dixon's 1868 picture of Prince and the Agapemone amounts to something rather far removed from our modern stereotype of the dangerous sect. Dixon presents, in effect, an image of a movement and a mentality in which anti-sensualism and pro-sensualism are so curiously intertwined that there is no pernicious contradiction between Prince's concubinage (or even public copulation) with Miss Patterson and his disciples' regime of celibacy, and no reason to deduce that Prince was habitually lascivious. Other evidence, including Prince's own doctrinal writings, which are fluent and full, confirms the strange interdependence of austerity and licence in his life and teaching. It is this which chiefly makes the Princeite episode intriguing in the present context.

An ex-Agapemonite recalled in 1905 that, apart from the separation of the sexes, life in the community 'was just like an ordinary country house'. The Agapemone was indeed luxurious and relaxing, but its rule of celibacy was by no means a vestige of monastic strictness in a climate of decadence. When Dixon visited he found a billiards table in the chapel, and was told that Sundays were consciously ignored by the Princeites. Such had been the spirit of the Agapemone right from its foundation: 'they profess merriment . . . they . . . delight in hockey, especially on Sunday . . . They . . . substitute feasting and enjoyment for privation and prayer'. 'Profess' is an important term here; the hedonism of the Agapemone was not an unthinking matter.[61]

At this date (1851) Prince was 40 years old, and only just over a decade away from his theological training at St David's College, Lampeter, and ordination. At Lampeter his intense commitment to the religious life had started, and he drew together in prayer-meetings his first small corps of disciples, the 'Lampeter Brethren'. The fervent and austere posture of this group within the world of Lampeter has been likened to that of the Wesleyans at Oxford a century earlier (and student life at Lampeter—a recent foundation which was widely regarded as a failure—was allegedly more luxurious and dissipated than Oxford itself). Prince did take a priggish stand on the 'levity' of his fellow-ordinands and their teachers: he records in his journal that when the 'drinking of wine and singing of songs' started at a dinner 'the Lord enabled Brother R. and myself to

[61] Montgomery 1962: 184; Dixon 1868: i. 236–7; *Illustrated London News* (29 Mar. 1851), 253.

leave', and this document, covering the years 1835–9, is full of anti-sensualism.[62]

But the great text which the Lampeter Brethren focused on together was the Song of Solomon, which Prince valued, it seems, both for its 'mystical quality' and for its 'vivid carnality'. In 1842, installed as curate at Charlinch but about to be forbidden to preach by the Bishop of Bath and Wells, Prince wrote,

the gospel does give liberty; but . . . it is a liberty in the *spirit*, not a freedom in the *flesh* . . . that liberty, however fair it may appear in other respects, is very much to be mistrusted, which does not keep the *flesh upon the cross* . . . the calls of God upon the Christian are sometimes so repugnant to his fallen nature . . . that they really do involve a very heavy cross . . . very grievous to the flesh.

Clearly there is a psychological preparation here, in the intense accent on 'flesh' versus 'spirit', and the gratuitous recognition of how 'fair' is fleshly liberty and how 'repugnant' fleshly restraint, for the embracing of a pro-sensual creed. This is a man with a strongly hedonistic temperament.

A theological route to pro-sensualism is also opened, such that no renunciation of Christian austerity is required. The line of thought is a recognized one in nineteenth-century sectarian belief, and is known as Perfectionism after the usage of some American groups. Ronald Knox has analysed it as 'an exaggeration of the Evangelical standpoint; of its conviction that salvation is something to be achieved here and now, that all things are made new to him who is reborn in Christ'. In Prince's lifetime the link was recognized between American Perfectionism and the Princeites, 'maintaining that there was no further necessity for prayer, and using their chapel as a luxurious drawing-room in token that the new life is a heavenly life of continual enjoyment and thanksgiving'.[63]

For Prince, with his very developed sense of 'the filthiness of the flesh', the logical step of Perfectionism was made significantly easier by certain doctrines about sacred history and about his own role in its unfolding. Again, the link back to the Evangelicalism of his Lampeter days is continuous. Though he remained discreet about his beliefs Prince was apparently convinced of a close link between his own

---

[62] *St David's College* 1851: 13; Price 1977: ch. 3; Prince 1859: 59–60, 97.

[63] McCormick 1962: 49; Dixon 1868: i. 257; Prince 1842: 129–32; R. A. Knox 1950: 566; Blunt 1874: 446.

person and the third element in the Trinity, the Holy Ghost, even
before he was ordained, and regarded his disciples in the Lampeter
Brethren as pentecostally endowed, like the apostles (there is some
slight evidence of 'speaking with tongues' at the Agapemone). The
title by which he was known in the Agapemone, 'Beloved', derives
from the Song of Solomon and was probably first adopted among the
Brethren, several of whom became founding members of the Spaxton
community.[64]

In the Princeite historical scheme the life of Christ had simply ini-
tiated one of a sequence of eras, and this particular era had closed in
the present. The one which had now opened was both the era of
Judgement, as opposed to that of the Gospel—and was thus actually
part of the eschatological process foreseen in Revelations—and also
the era of the Holy Ghost (Prince equivocated on his role here, sometimes insisting that he was only John the Baptist to the Holy Ghost's
Christ, but often writing of himself or allowing himself to be presented as the very incarnation of the Holy Ghost). In the new era,
above all, the *flesh* of Man would be redeemed, the era of Christ and
the Gospel having redeemed only his spirit. The possibilities for a
Perfectionist account of Prince's own actions, given these ideas, are
obvious enough. In so far as the flesh is only sinful because it has yet
to be redeemed, and he is the agent, or at least the instrument of the
agent, of that redemption, his 'liberty' in the flesh is indeed as 'fair' as
it seems. Prince taught that the flesh which had taken Miss Patterson
was 'spiritual', though the flesh it took—hers—was 'natural' (a
Swedenborgian sounding distinction, incidentally).[65]

There was another avenue to Perfectionism in Prince's thought,
which potentially made its moral freedoms available to everyone in
the Agapemone. Prince placed an interesting emphasis, somewhat in
the manner of Eastern religions, on the loss of the sense of personal
identity as part of the process of redemption. It was indeed not difficult to equate a renunciation of the 'consciousness of self' with the
traditional experience of Nonconformist conversion (and Prince continued to teach, like any good Methodist, 'the precious doctrine of
justification by faith without the works of the law'). The unusual
move was to relate the consciousness of self also to the knowledge of
good and evil acquired by Adam and Eve at the Fall: 'what *is* the
knowledge of good and evil? It is the *knowledge, or consciousness, of*

---

[64] Deck 1845: 8–16; Rees 1846: 7; *Illustrated London News* (29 Mar. 1851), 253.
[65] Prince 1856–9: 'Voice the Fourth', p. 29.

self; 'what was this new thing they were so concerned to hide? It was SELF'. There is a circularity here which seems sinister to the unsympathetic eye. If one is not conscious of sin there is no sin to be conscious of. Certainly Princeites attached importance to their own lack of dismay, or at least surmounting of dismay, at whatever it was they witnessed being transacted between Prince and Miss Patterson, which is why they stressed its public nature. Years later Dixon was assured by the leading disciples that they could see no wrong in what had happened.[66]

Granted that the Princeites were Perfectionists, enjoying the things of the flesh in recognition of the conquest of sin, why was their pro-sensualism not more uniform in its operation? On the face of it, sex was in a special category at the Agapemone, being the one kind of pleasure from which abstention was still urged. It is difficult to make a judgement about this, however. The other 'country house' pleasures of the Agapemone—the billiards, the hockey, the wine and food—may have been quite moderately and decorously indulged in. This was a sect which believed, like others of its kind, that there was no more death for its members, and that the rest of the world would soon be destroyed in the making of a new heaven and a new earth. There may have been a considerable serenity, and little avidity, in the way they took their pleasures: Dixon's account gives something of this impression. As for sexuality, the Agapemone was 'The Abode of Love', and affectionateness was presumably the keynote of all relationships in the community: the mood was probably pro-celibate, but not unerotic. Prince himself, on the question of marriage, taught a qualified morality. Abstention becomes more desirable as Man rises spiritually; in scripture, accordingly, the injunction to marry in *Genesis* gives way progressively to Paul's advice against marriage. There is some affinity here with secular ideas of the time on the decay of the sexual impulse.[67]

What of the apparent disparity between the sexual code of the disciples and that of their leader—who in so undisguised and thoroughgoing a fashion took a mistress? The scarcely credible truth here seems to be that Prince did believe his action to be spiritual rather than physical. Miss Patterson's pregnancy was an event deemed impossible, and it gave a blow to the sect's theology from

---

[66] Prince 1900: 140, 1856–9: 'Voice the Fifth', p. 24; Dixon 1868: i. 317; *Bridgwater Times* (21 Aug. 1856), 3.

[67] Prince 1900: 296–7.

which it was only just able to recover. Prince himself was said to have been 'overthrown by grief and shame' to find that his natural body had not, after all, passed away. For a year, perhaps, the sect was cast into doubt and despondency, and the mood of that time was still vivid thirty years later to a Princeite outside the Agapemone. An official explanation was eventually developed and duly patched into the doctrines of the sect. Miss Patterson's child was 'Satan's offspring', Prince's (and perhaps Miss Patterson's) 'conviction of total ruin' was also Satan's work, and the whole episode had been providentially designed to 'bring to light the Wicked one'. There is a renewed emphasis in the Princeite literature on the satanic nature of our fallen flesh (the selection from Prince's Lampeter journal is issued at this date), and it seems likely that from this time on *all* members of the community practised chastity.[68]

The Princeite movement sprang out of Evangelical Anglicanism, thrived on middle-class support, devised a monastic rule curiously like country-house life, and attracted more published comment than any other small sect in the middle years of the century. This may distract attention from the many important respects in which it was out of tune with the spirit of the times, both generally and on sexual questions. The Princeites were fanatically committed to the 'Beloved', Henry James Prince, and there is an astonishing unanimity in all the utterances which issued from the Agapemone. They seem to have believed quite literally in Prince's supreme historical significance and in the imminent end of the world. They tended strongly to favour a celibate life for the elect. This aspect of their doctrine (though it applied also to the Princeite clergy and laity outside the Agapemone) may have been obscured for the general public by the perception that the Princeites were 'free lovers', but attacks on matrimonial sex from the libertine side were just as unpopular in mid-Victorian England as those which came from the advocates of celibacy. The sect's great crisis blew up at the time of the publication of George Drysdale's *Elements of Social Science*, and the brief flurry of scandal which followed may be seen as part of the second wave of liberationist controversy in the century, just as the bigger clash over Swedenborg's *Conjugal Love* is part of the first.

On the other hand, there was a definite modernity to Princeism, in its day. No monastery remotely like the Agapemone had been known

[68] Dixon 1868: i. 250, 325; W. Fox 1886: 12; Prince 1900: 177, 1858: 167–83; G. R. Thomas and S. Starkey 1862: 8–10, 135–6, 249. But cf. Mander 1976: 131.

in Christianity, and some of the sect's beliefs and attitudes were unmistakably of their time. The doctrine of celibacy as an easily borne condition appropriate for the elect at the end of history has more in common with utopian anti-sensualism than with the traditional asceticisms of the church (Prince actually deplored the extreme austerities of early Christianity).[69] The emphasis on the sense of guilt as itself a kind of taint has a ring of the Romanticism of Blake and Shelley to it, and even of the Godwinian analysis of the despotism of reactionary moral opinion. As with the Swedenborgians and the Irvingites there is, in effect, an abandonment of ancient ideas of damnation and punishment. The Princeites believed in sin and the fall, but also taught that it was sufficient to transfer oneself to comfortable premises in Somerset in order to be saved at the judgement, because this judgement was in fact already under way.

The sexual teachings of nineteenth-century sectarian groups could be investigated almost indefinitely. Some of these groups will be partly catered for shortly in the discussion of large denominational units, such as Evangelicalism and Methodism. Others were vanishingly small: such as the Corkelers, a sect founded in 1850 and confined to the Surrey–Sussex border, who treated marriage as a regrettable fact of life, and the slightly later Walworth Jumpers or Girlingites—who did indeed vanish, as they were entirely celibate.[70]

Yet other religious factions with a strong sexual element in their teaching shaded into political movements, and as a result will make an appearance later in this book. Various descendant groups in the Southcottian church, following Joanna Southcott's death in 1814, are the most salient instances here. The movement as a whole was very large, numbering at least 20,000 followers, but perhaps even ten times this figure, and after 1814 it dispersed under a number of different leaderships. Two of these figures, Zion Ward and James 'Shepherd' Smith, had strong views about sex, in different veins, and both of them were caught up in the Owenite movement. E. P. Thompson says that large numbers of Southcottians converted to Mormonism: their story therefore lies mainly on the other side of the Atlantic, since so many Mormon recruits emigrated to America.

George Turner, an elderly Leeds merchant and ex-follower of Richard Brothers, who was one of the first to make a bid as Joanna

---

[69] Prince 1900: 123–5.

[70] Montgomery 1962: 105–20, 209–18; D. Hardy 1979: 137–48; Davies 1873: 89–99; Ritchie 1870: 307–13; Ross 1887: ii. 3–16; Housman 1937: 50–1, 116–17.

Southcott's successor, taught that the apocalypse must be heralded by his 'marriage' to some 1,556 female Southcottians. This project he seems to have launched in 1817, and attempted to achieve by means of country-wide individual ceremonies and a collective ceremony in London in 1820. In the interval he was confined in the Retreat, the York madhouse, for allegedly treasonous apocalyptic predictions. It is apparent from his various writings at the time that the physical content of these 'marriages'—whether the wives should be 'wedded and bedded'—was the subject of a good deal of heart-searching and dissension for Turner and his disciples. The problem was solved by Turner's death in 1821.[71]

The figure of Turner is a reminder that there need not be a link between an individual's striking sexual code and his or her sectarian affiliation, however pronounced—or no link of importance. Some individuals, like Turner, were on the verge of insanity. Others were no doubt just hypocrites. The notion of the outwardly austere minister who is secretly a sexual predator was, of course, a cliché, and it is impossible to assess the many anecdotes of this sort to be found, for example, in the memoirs of the ex-Methodist bookseller James Lackington. But some figures do seem to stand out as pretty raw charlatans. It is hard to see how else the astonishingly pious and moralistic letters the publisher Joseph Stockdale wrote from prison around 1830 are to be reconciled with his list, which with great consistency was made up of books and periodicals about sex: sexual medicine, sexual memoirs, *risqué* novels, and *crim. con.* proceedings.

And some religious individuals were probably just in a state of painful contradiction between their beliefs and their sexual inclinations. This seems to have been the case with John Church, the homosexual minister of the Surrey Tabernacle. His is another story of a sexually transgressive religious leader managing to keep the loyalty of his congregation, in this instance after he had been imprisoned between 1817 and 1819 for a homosexual offence. Church's sermons of farewell and return, and his letters to his followers from prison, do not directly admit any betrayal of standards (this would have been especially difficult in relation to homosexuality), but they do register a sense of abasement, and a struggle to come to terms with this. It is interesting to see Church drawn by, but on the whole rejecting, the

[71] Balleine 1956: 75–82; J. E. Smith 1833*b*: 95; J. K. Hopkins 1982: 34; E. P. Thompson 1968 p. 882; J. F. C. Harrison 1979: 119–21, 188–9; A. M. Smith 1987; Turner 1819*a*: 110–11, 139, 1819*b*: 54–6, 65, 70, 1820*a*: 14–15, 35, 1820*b*: 54–5.

straightforward antinomian solution to his moral predicament: that the 'elect person' cannot be damned. He takes comfort from the thought of his salvation, but it is evident, I think, that he is not comfortable: 'there may be a bias, a propensity to evil, but you can never be a carnal man again—you may look like one, feel like one, and suffer like one; but though this may grieve you, it shall never be accomplished in the way Satan desires'. Church's enemies accused him of outright antinomianism, but this seems to be unfounded.[72] It was a common slur on the Calvinistic sects, but probably never a fair one, that their members used their election as a shield for all sorts of vileness. Here was one instance in which the remarkable elasticity of the Christian system was not exploited in the sexual codes of nineteenth-century believers.

III

Classic Moralism

The fact that in the early nineteenth century Methodists and other dissenting sects were habitually accused of fostering sexual immorality is worth pondering. This cannot be a case of Evangelicalism biting the hand that fed it, as is sometimes argued of high-Victorian attacks on Evangelicals, because the timing is wrong. The idea is a familiar one that the Victorians were infected by the 'seriousness' of Evangelicalism but not by its literal doctrine, and that writers who owed much, unconsciously, to Wilberforce and Hannah More were thus left free to satirize a Mr Stiggins or a Mrs Pardiggle. I shall consider the plausibility of this argument shortly, but it is obviously inapplicable to, say, Leigh Hunt's attacks on Methodism in the first decade of the century, which are rich in jibes about love-feasts and about the sexual imagery of methodistic devotional discourse.[73] Moralistic criticism of the immorality of the religious is a very familiar and venerable theme. Po-faced lechery goes back at least to Angelo in *Measure for Measure*, and in a sense to Chaucer's Pardoner. It starts to look as if denominational religion, of whatever sort, had little to do with this particular phase of the period's sexual moralism.

How much active religion was there in nineteenth-century England as a whole, and in what forms? Two kinds of comparative

---

[72] Bell 1813: 29; Church 1817: 13, 1818: 40, 62, 97, 108, 166.
[73] Hunt 1809: 56–8, 90–2.

information are available: estimates of affiliation made by modern historians using the records of the individual churches, and the survey in the 1851 census of church attendance on 30 March of that year.[74] The latter is only a snapshot of churchgoing at one date, while the former approach traces the fortunes of the churches through time. It is clear from the census that the recorded affiliations used by historians (whether measured by membership of a dissenting church or by attendance at Anglican and Episcopalian communion) are generally much lower than the levels of church-worship: only the Roman Catholics showed an opposite pattern, attending services in 1851 in greatly fewer numbers than the total baptized into the church.

So a certain amount can be said quantitatively about the life of religions in nineteenth-century England, but not much about the religious life. The 1851 figures are recognized to be only a rough indication of how many individuals went to church on the Sunday in question, but they suggest that about 40 per cent of all English and Welsh men, women, and children (i.e. just over 7 million individuals) attended a service. The number of Catholics counted, as I have mentioned, was well below the potential level, amounting to about a quarter of a million out of an estimated 850,000 or so baptized persons. In the Protestant sector, roughly equal numbers seem to have attended Anglican establishments on the one hand, and all Nonconformist establishments put together on the other. Given the small number of Catholic attendances, this means that the established church and Nonconformity split the churchgoing public approximately in half, each taking 20 per cent of the whole population of England and Wales. Not unexpectedly, the discrepancy between the numbers affiliated to the Protestant churches (in the sense defined above) and the numbers attending services in 1851 is much greater in the case of the Church of England than it is with the Nonconformist denominations. Very broadly, only about a quarter of the Anglican attenders were also in the habit of taking communion on Easter Day, but about three-quarters of the Nonconformist attenders were known members of the congregations they were counted in, or of another congregation in the same sect.

The pattern of lukewarm Anglican allegiance makes a not very strong established church look even weaker. But the other kind of statistical picture which can be drawn, representing the fortunes of the

[74] Gilbert 1976: ch. 2; Currie, Gilbert, and Horsley 1977: 132–66; E. J. Evans 1983: 422–7.

churches across time, renders the Anglican performance more impressive. The hard core of Church of England worshippers (those communicating at Easter) held up and actually slightly increased in the Victorian period as a proportion of the total population, after at least a century of decline. The number is always small, bottoming out at about 4 per cent of the population in 1830, and never rising above 5, but it is a real increase, to be contrasted with the fortunes of the Nonconformist churches in the same period. The main denominations in this sector—the Methodists, the Baptists, and the Congregationalists—all peaked as a proportion of the population around 1850.

England and Wales in the nineteenth century did not form a very religiously active community. But as far as it went its religious allegiance, by a narrow but secure and increasing margin, was Anglican (obviously the facts are different area by area, and I shall mention some local variations later). This would be a purely numerical point, however, and uninteresting for the understanding of religious attitudes at the time, if there were large numbers of Anglicans doctrinally in sympathy with Nonconformity. I have already used the term 'Evangelical' to cover both dissenting groups and an element within the Church of England: Sydney Smith, in his celebrated *Edinburgh Review* articles of 1808–9, called the same people 'Methodists', and this was a standard usage. The scale of the Evangelical/Methodist fraction in the Church of England is particularly relevant, of course, to the moral culture of the period.

There is no formal criterion of 'Evangelical' sympathies in a clergyman, and it has naturally proved difficult to arrive at hard numbers at the clerical level. It has been suggested that as many as a tenth of Anglican ministers in 1800 were Evangelicals, that is, about 1,400 individuals, but according to another estimate the number may have been only about a third of this. Even then, the expansion would be extremely impressive, since a decade earlier the total of such individuals was perhaps no more than sixty. By 1830 there was certainly a further substantial increase: to somewhere between a quarter and an eighth of the whole ministry.[75]

At the mid-century, in W. J. Conybeare's widely read analysis of the varieties of Anglicanism, the strictly methodistical end of the Evangelical party was labelled the 'Recordite' faction, after the newspaper which represented its views, and was said to be on the decline.

[75] Hylson-Smith 1988: 68; Gilbert 1976: 73; F. K. Brown 1961: 2, 9; Bebbington 1989: 106–7; E. Jay 1979: 23–32.

The circulation of the *Record* was about 3,500 (there were 16,000 Anglican clergy by 1850), though it must be said that it was the only Church of England newspaper of any sort available until 1843, and probably had readers who did not like its views. However, among those alienated by the *Record* were many Evangelicals of a less strict, and politically more liberal, cast. Those linked to the Clapham Sect, in particular, were hostile to the Recordites. Conybeare judged that the Evangelical party at large was still prospering. Once the Evangelicals had a firm toehold at the top of the Anglican hierarchy, which they achieved in the 1830s and then enlarged in the 1850s, the level of ordinands of a similar persuasion was bound to keep up, though it has been suggested that there was a loss of actual expansive energy when Wilberforce and Simeon left the scene in the 1830s. Certainly there was a loss of Anglican Evangelical momentum from the mid-1860s, quite unforeseen at the start of Victoria's reign, which baffled the participants at the time and has since baffled historians.[76]

However, Evangelicals were evidently a substantial group among the ministers of the Church of England. How might this have translated in terms of their influence on the laity? By definition, an Evangelical has a special commitment to Christian teaching, and one may be sure that the numbers of Evangelical clergy were more than matched by the scale of their activity in English parochial life. But with what effect? Interestingly, Evangelicals themselves tended to feel that their real impact in particular communities and at large was small compared with what they hoped and others perceived. This was partly a reflex from insatiable missionary zeal, but Charles Simeon specifically denied that more than 1 per cent of the population of Cambridge were committed Evangelicals.[77] A distinction should be drawn between the small number of Recordite ministers, with fervent congregations which resembled dissenting chapels, and the less zealous majority. The latter would certainly have offered a more high-minded, industrious, and spiritual image of the clerical life than had been known in the church a hundred years earlier, but one modern expert on nineteenth-century religion and society has judged that the impact on the lay public may not have been commensurate: 'Anglicanism became for a time a more ubiquitous religious-cultural

---

[76] Conybeare 1853; Ellegard 1971; Altholz 1987; Hylson-Smith 1988: 50, 68, 96, 102; Englander 1988; Newsome 1966: 8–9; Lewis 1986: 16–17 and *passim*.

[77] *Christian Observer*, 10 (1811), 484–92; Halevy 1924: 384.

phenomenon in English society, but, arguably, it did so only by becoming a less pervasive one.'[78]

In particular, the revitalized Anglican ministry was losing, or at least not increasing, its power to influence middle- and upper-class opinion. Although clergymen at the mid-century were younger than at any previous period of the church's history, and were motivated by an enhanced sense of the dignity of the pastoral office, much of their effort went into urban, working-class evangelism. Also, the social rank of Anglican clergymen, and the standing of priesthood as a career, declined through the Victorian period.[79] As those who joined its numbers became more unworldly they may also have seemed less relevant, except for those parishioners who could respond to the new and arduous standard for the personal relevance of religion which had been set by Wilberforce. On the other hand, Evangelicalism was in the air. The country was awash with millions of Evangelical tracts; until the 1850s, at least, religious publications were edging up to a third of the sales of all titles; and there was a massive Evangelical effort in, and on behalf of, the provision of Christian texts and teaching. Full-time lay workers in the various urban missions bulked comparably to ministers of religion: in a ratio of about one-to-two in London at the mid-century.[80]

The direct targets of these ventures were mostly the working class and foreign countries, but the spirit of the enterprise could not be ignored, to say the least, by the educated English. And it is arguable that social changes amplified the influence of the Evangelical attitude out of proportion to the number of ordinary churchgoing Anglicans who had made a conscious commitment to Evangelicalism. The idea was familiar by the late nineteenth century that modern respectability represented the diffusion upwards and downwards of the moral-religious code of the lower bourgeoisie (Anglican Evangelical and/or Nonconformist). With the advent, at the time of the Great War, of the 'Victorian' model of nineteenth-century respectability comes the suggestion that 1832 represented the rise of the Evangelical English middle classes not only to political but also to moral dominance.[81]

Nevertheless, of the three million or so English men and women who did attend services in the Church of England on 30 March 1851,

[78] Gilbert 1976: 134.
[79] Haig 1984: 4, 14–15, 29–32; Gilbert 1976: 133–4; Heeney 1976: 126.
[80] Lewis 1986: 120; P. Scott 1973.
[81] G. W. E. Russell 1899: 64; Traill 1891; Christie 1927: 75–6.

very many, probably the majority, were unconverted by the Evangelical revival, or only lightly touched by it. There were also hundreds of thousands of individuals who did not attend Anglican services but who were christened Anglicans, and themselves married in the church and christened their children into it. Nominal Anglicanism in this sense must in fact have greatly exceeded committed Anglicanism as measured by the 1851 census. Some of it was nothing but nominal. Frederic Harrison, with whom this chapter started, was an Anglican in the most extended sense: he and his bride, like some of their like-minded associates, simply gave in to their families' desire for a conventional church marriage ceremony.[82]

Harrison, the positivist, the trade-union propagandist, the promoter of working-class education, would no doubt have justified his belief in sexual continence mainly in terms such as those used by progressives in attacking Malthus. But he was also aware that his views touched hands with Christian opinion on sex. Many luke-warm Anglicans at a higher temperature than Harrison, it may be guessed, would have appealed to the Bible if they were challenged to justify their sexual moralism, unaware that Christianity's anti-sensualism was a matter of history rather than text. But the stock of holy writ was sinking steadily through the nineteenth century, and that of history rising. Leigh Hunt, with whom this section started, had a religious creed which he called 'Christianism': the name grants Christ a priority among religious teachers, but is also supposed to convey the thought that there were other authorities in the ancient world, such as Plato, Epictetus, and Marcus Aurelius, whose utterances can be homogenized with those of Christ.[83]

Most of the prohibitions in the English nineteenth-century sexual code were venerable ones. Male and female masturbation, unconstrained liaisons by either sex, non-marital intercourse by women, and high coital rates in marriage have been deplored, repressed and even punished in the majority of known human societies. We tend to perceive the nineteenth-century code as a novelty, but this was not at all how it seemed to its followers. To them the nineteenth century appeared to be, at most, implementing with greater success than previous epochs an age-old set of values which were the only conceivable morality about sex. Thomas Buckle explicitly argued that moral

---

[82] Vogeler 1984: 91.   [83] Hunt 1853: 150–3.

standards did not change with the advance of civilization, only mankind's capacity to live up to them.

Buckle represents moral traditionalism given a thoroughly progressive turn: the survival of values in history is interesting for what this allows us to prognosticate about the future. This is a variant of progressive anti-sensualism of great importance in the period. It differs from that of Mill, for example, in that the chronology of progress is much larger and hazier. Locally, as Robert Chambers granted in the very influential *Vestiges of Creation*, it may even appear that the sexual standards of humanity are deteriorating, but we can none the less be confident that, over a sufficiently long temporal span, 'as reason and the higher sentiments of man's nature increase in force, this passion is put under better regulation'.[84] Also, the bigger scale means that periods in the past may have exhibited an admirable degree of moral advance—something which the political beliefs of most radicals, before William Morris at least, ruled out. It was a commonplace of this way of thinking that the chivalric code of the Middle Ages represented a leap forward in sexual standards, a piece of historiography which Mill had no patience with.

This 'aspirational anti-sensualism', as I would term it, was particularly influential on the serious imaginative literature of the period. This does mean that there is a danger of overestimating its role, undoubtedly significant though it was. In our textual experience of Victorian culture the great literary authors naturally tend to loom large. Consequently the account of Victorian attitudes towards sexuality given in chapter 13 of Walter Houghton's classic *The Victorian Frame of Mind*, for example, implies that aspirational anti-sensualism was the main motor of the sexual moralism of the day. But the compounded optimism and conservatism of Tennyson or Ruskin is far from being the whole story about the contribution of the forward-looking mentality, or of the backward-looking, in the formation of Victorian sexual standards.

There was, indeed, a vigorous survival of starkly conservative attitudes about sexual morality, at odds with sexual licence, certainly, but in some ways more antagonistic to sectors of opinion with whom it might be expected to have made common cause. This classic moralism was in conflict with revivalist religion on some of the latter's sexual beliefs—overt doctrines, these, and not the secret immoralities of

---

[84] Chambers 1844: 368.

Methodism. It favoured Malthusian restraint, for example, while Michael Sadler was clamouring for early and prolific marriage (Morse Peckham has astutely characterized Evangelicalism as a 'counter culture'; he has failed to draw the corollary that it was in some way less restrictive in its moral views than the establishment culture).[85] Classic moralism probably thought the Swedenborgians carried the celebration of matrimonial love too far. Above all, it clashed with Evangelicalism on the treatment of prostitution, as I shall describe. But classic moralism was fast failing in strength by the end of the period covered in this book. Its rearguard action, and defeat, on the licensing of prostitutes was the last conflict on a sexual issue in which it had a major role. Thereafter the main anti-sensual groups are shiny new ones advocating 'social purity' or feminist celibacy.

The key to this classic moralism is that it was opposed to the theoretical. It had a distaste for systematic and speculative inquiry about sex; it operated from the given. This affected its outer and inner character in several ways. It was a creed expressed more in private precept than in essays and books, which is why its strength was greater than the published record would suggest. It was not too troubled about consistency, and could embrace a variety of tones and attitudes without discomfort. It liked to think it had its feet on the ground, and had derived its beliefs from an honest picture of human nature and the world. It also took as among the given a stock of inherited beliefs which it did not choose to enquire into.

Classic moralism, despite being the establishment code of the time (if there is such a thing), kept an anti-establishment ring. It favoured ephemeral, witty, and destructive modes of expression (satire has at all periods been a major vehicle for its values). It emerged occasionally in a serious and granitic form, but was always liable to a kind of leavening and disrobing. Two serious statements of the classic moralist position on sex can be taken as forming an entrance to its history in the nineteenth century. One author, William Paley, writes from the heart of Anglicanism, and the other, Dugald Stewart, out of the eighteenth-century deist tradition, and their pronouncements are conveniently simultaneous.

In both cases sexual doctrines take up short sections of longer enterprises: Paley's *The Principles of Moral and Political Philosophy* of 1785, and Stewart's *Lectures on Political Economy* (which probably

---

[85] Peckham 1975.

also date from 1785, delivered as part of Stewart's celebrated lectures as professor of moral philosophy at Edinburgh, though they were not published until after his death).[86] In both texts the treatment of sexuality is terse, but entirely confident. There is an almost complete identity of views between the two men: celibacy is an evil, but so are all expressions of sexuality other than permanent monogamy; prostitution, concubinage, and divorce are all vicious; if they were encouraged they would be exploited mainly by men, who are much less chaste and modest than women.

Paley, the theologian, does not much appeal to theology in making these pronouncements, and Stewart, the philosopher, not at all to philosophy. For Paley the facts of society and human nature are a sufficient warrant; Stewart's chief authority is Greek and Roman precept and anecdote. Indeed what passes for argument in Dugald Stewart, supposedly the heir to Hume and Thomas Reid, is so piffling that the great enthusiasm aroused by his lectures now seems incomprehensible. But he was an inspiration and influence for an extraordinarily illustrious group of nineteenth-century men, including Sydney Smith, Lord John Russell, Palmerston, and Henry Brougham. His personal moral example, of extreme uprightness, seems to have been an important reinforcement for the thin moralizing of his lectures.[87]

Dugald Stewart was above all admired by the spirited, slashing literary journalists—such as Gifford, Jeffrey, Lockhart, and Wilson—who were so distinctive of Edinburgh culture early in the century. This indicates how classic moralism found a natural home in witty aggression, and often in wittily aggressive coteries. In this sense it was the moral face of illiberal conservatism, though there were plenty of unreflectingly moralistic Whigs who must also be accounted disciples. The *Anti-Jacobin Review*, for example, took a classic moralist line on the few occasions when it touched on sexual issues. It is mainly because these references are scarce that the *Anti-Jacobin* will not figure in my brief survey of some classic moralist productions. But this school of anti-sensualism did tend anyway to be fragmentary, organizing itself not into national movements but into clubbish units. This was another result, in fact, of its antipathy to theory and system.

Regrettably little is known about the personnel behind the *Scourge*,

---

[86] W. Paley 1785: 243–59; D. Stewart 1855: 68–96.   [87] Jaeger 1956: 82–9.

a conservative though Whiggish monthly dedicated to the exposure of 'imposture and folly' which ran from 1811 to 1815.[88] It was a middle-class vehicle of moral satire in contrast to the more aristocratic *Pic-Nic* and *Satirist*—though the latter, in particular, is very comparable as a voice for classic moralism. The *Scourge* offered much sparkling writing and a valuable hoard of information on contemporary sexual culture. George Cruikshank executed plates for it, so the *Scourge* is linked to the main English tradition of graphic satire. Because the magazine was a blithely energetic affair, and just as quick to become indignant about contemporary vice as about contemporary puritanism, it ran fairly frankly into self-contradiction. On one hand the 'common-places' are right, says the *Scourge*, to the effect that 'the present age surpasses in degree and variety of wickedness all that have preceded it; that the sum of national virtue and domestic happiness is considerably diminished'. Four years later, these commonplaces have become 'the incessant lamentations of . . . modern Jeremiahs', who claim the world is 'approaching rapidly to the very *acme*, and *ne plus ultra* of wickedness', while the *Scourge* doubts if conjugal infidelity, for example, is any more rife than in the past.

Logically the *Scourge*'s many attacks on the Vice Society are only a little more happy. This morally interfering body (which I discuss in the first section of the next chapter) is nailed moralistically, by the *Scourge*, for having persecuted the wrong people: minor peddlers of obscene prints rather than prostitutes, for example (such comment, however confused, is useful in shedding light on the work of this underrecorded organization). The *Scourge* printed complaints about London's prostitutes as outraged as any that can be found in the period, but in another mood, and in the spirit of classic moralism right through the century, argued at some length for recognition and regulation of prostitution as an unfortunate necessity: 'this position no person of common sense and discernment will attempt to contravene'. Indeed, the Vice Society's alleged mistake is now precisely that it *does* try to restrict prostitution: the result of its efforts will be an increase in masturbation, adultery, and homosexuality.[89]

Some of these tergiversations may be due to a change of editorial control, but the *Scourge*'s readership was presumably of a piece, and contradictions of this sort are endemic to the classic moralist position.

[88] See Sullivan 1983: 388–91.
[89] *Scourge*, 1 (1811), 313–16, 377–8, 2 (1811), 22, 137–8, 3 (1812), 35, 4 (1812), 76–9, 9 (1815), 44, 91, 10 (1815), 256.

The Vice Society, for its part, no doubt regarded the *Scourge* as obscene, and there would be some real justice in the accusation—for all the magazine's protestations of attachment to 'the family parlour of the middle classes' where 'morality holds her court'. There is always a tactical problem for satirists that if they are too knowing about vice their own probity may seem doubtful, but the *Scourge* rather tended to embrace this effect, shading from satire on Evangelical prudishness into dirty jokes of its own manufacture:

> every unmarried female, above the age of fourteen, detected in assisting a male child, to perform the urinary office, should be committed to the county gaol ... Of the equine race, mares and geldings alone should be permitted to approach within three miles of any populous town; and every object that bears a *Phallic* outline, should be banished from our dwellings, and razed from out streets ... rolling pins should be burnt without mercy, and pokers be driven from the parlour stove.

An extreme example of this vein is 'we suppose . . . committees will be appointed to ascertain how many inches make a yard'.

Obscene puns of this sort make the magazine's claim that 'chastity is in the mind, and not in the word' no more than a Benny Hill leer. Sometimes, indeed, the *Scourge* seems to be attacking not so much euphemism of language, as a kind of euphemism in contemporary sexual self-awareness, and one can see that the no-nonsense, anti-idealistic spirit of classic moralism leads easily to ribaldry. A parodic piece on 'Specimens and proposals for a new dictionary' makes fun less of prudishness than of sexual high-mindedness, which it reinterprets with extreme reductiveness: '*Love.*—A contagious disease ... It may be cured by a very simple process if the remedy is *in hand*'.[90]

'Yard', in its sexual meaning, is frankly ribald, while 'the remedy . . . *in hand*' is a dirty innuendo, a post-Sternean affectation of verbal innocence. For the *Scourge*'s Victorian descendants even innuendo became less available for the expression of a reductive or sceptical attitude towards sexuality, a change which, as I see it, was against the grain of this brand of anti-sensualism (in other words, classic moralism is not one of the forces making for respectability in the written word). However, importantly, the underlying reductiveness about sex remained intrinsic to the classic moralism of the Victorian years—and was perhaps particularly current in an age which offered so many provoking examples of high-mindedness about sexual

[90] *Scourge*, 1 (1811), 166, 2 (1811), 23, 207, 8 (1814), 200–4.

relations. It also found an increasingly scientific atmosphere to foster its growth.

The Vice Society did indeed prosecute a Victorian magazine which like the *Scourge* sought to expose imposture and folly, namely, Renton Nicholson's the *Town*. 'Baron' Renton Nicholson is a conspicuous but puzzling figure, a force in sexual culture right through the first half of Victoria's reign whose standpoint is hard to assess. On the face of it Nicholson's chief enterprises—the *Town*, Cremorne (the pleasure gardens in Chelsea noted chiefly as a place for dancing and other evening entertainments, which Nicholson owned from about 1845 to 1861), and the Coal Hole tavern on the Strand—show him to be an entrepreneur in the sexual domain, a man with an eye for commercial possibilities in the not respectable but not thoroughly disreputable sector of London's recreational life. He also edited the short-lived Whig magazine the *Crown*, which may have been a perfectly cynical venture or a serious expression of Nicholson's own political loyalties, and had a stake in the no less abortive *Illustrated London Life*.

He was above all famous for the 'Judge and Jury' show as run in the Coal Hole—and his sobriquet of 'Baron' derives from his leading role in the Judge and Jury performances. This was a form of satirical entertainment dating back to the 1780s, and in London at least there were evidently several other less celebrated versions on offer. Nicholson himself had first run Judge and Jury nights in premises in Bow Street, and took the show on provincial tour, but it was at the Coal Hole, opposite the great Evangelical centre at Exeter Hall, that the Judge and Jury show became an institution which everyone who lived in London, or who came to London from the provinces and abroad, seems to have known about. It managed to last for ten years or so after Nicholson's death in 1861, but was increasingly in trouble with the law; it is last heard of in Leicester Square.[91]

The many surviving descriptions of the Judge and Jury show agree that the comic legal framework was mainly used for fictitious or semi-fictitious divorce proceedings (or less commonly such things as breach-of-promise suits), though there are also references to political and cultural satire (with the 'social evil' as one target) being mediated

---

[91] Hibbert 1916: 12–14; Fryer 1968: 234; Pearl 1955: 180–7; Rendle 1919: 194–8; Sims 1917: 106; Vizetelly 1893: i. 168–70; *London by Night* ?1857: 44; Greenwood 1874: 99–106; S. T. Taylor 1927: 10; Fiske 1869: 173–4; Nicholson 1965: 282–7, 298, 311, 327; Wagner 1925: 161; Boase 1893.

by the device of the trial. Whatever the material, there was obviously a lot of dirty humour, with transvestism an important theme, and *poses plastiques*—stationary figures posing in tight, flesh-coloured costumes—were a regular ancillary feature. There is the usual obscurity, due to the prejudices of observers, about the actual level of obscenity in the Coal Hole. Nicholson's remarkable wit, and hence a certain cerebral element in the show, comes through clearly enough. In physical terms the venue was quite elegant, especially by the late 1850s, offering many creature comforts, and in this sense it had something of the ambiguity of the sister-operation at Cremorne. The 'Coal Hole' was obviously a comically misleading name, designed to give a *frisson* of naughtiness with its mixture of slummy, grimy, and hellish associations. Nicholson certainly exploited an antinomian vein in his publicity, consistently drawing attention to the proximity of Exeter Hall, and even handing out advertising leaflets to men in the West End in the form of parodies of the leaflets distributed to prostitutes by the Midnight Meetings movement.[92]

Before 1857 satirical presentations of divorce proceedings were by definition humour about the rich and great, with a potential political edge. The fact that Nicholson's Judge and Jury show, as far as can be gathered, was much the same affair before and after 1857 may imply that smut rather than politics was always his main commodity. In so far as Nicholson has ever been claimed for political activism it is as a radical, a survivor from the tradition of ribald anti-establishment satire of the early years of the century.[93] The best evidence about his stance is the magazine already mentioned, the *Town*. This 2*d*. weekly ran for four years from 1837. It is a puzzling production, with perhaps genuinely confused aims, which also exhibits the framework of a more coherent attitude, along classic moralistic lines, than might appear.[94]

Its staple is London's sexuality, and one cannot doubt that its readers liked it for the sexual facts and possibilities about the capital which it opened up for them in its descriptive pieces, gossip, and advertising (Nicholson called it 'a weekly necessity with the fast

---

[92] C. Scott 1896, 1897: 82; Texier 1851: 39; Fontane 1939: 114; Ritchie 1857: 76–82; Tuckerman 1854: 113–15; W. E. Adams 1903: ii. 310; Acton 1857*b*: 103–4; 1852 advertisement in Bodleian Library copy of Chancellor 1912; Doré and Jerrold 1872: 168–9; Yates 1884: i. 166; *Magdalen's Friend*, 2 (1860), 162; Nicholson 1965: 285–7; *Fast Life* 1859: 162–5; *Paul Pry* (25 Oct. 1856), 1–2; *Illustrated London Life*, passim (Judge and Jury announcements; Enault 1876: 328–9).

[93] McCalman 1988: 225.   [94] See Blanchard 1891: i. 257–8.

community').[95] Sometimes its protestations that all is done in the interests of exposing vice may strike one as the most tired and perfunctory hypocrisy, quickly shown up for what it is by some crude *double entendre* or smutty anecdote. But the *Town* does also quite regularly, and at some length, offer ideological statements or implied statements which are gratuitous, and over and above anything required by a pretence that it is an agent of morality. The positions which emerge here are those of classic moralism, traditionalist but self-approvingly expedient and realistic, and one may speculate that its ribaldry was received as a manifestation of the down-to-earth spirit of good old-fashioned sexual values (while of course offering also the pleasures of titillation).

The *Town* seems *not* to have been radical in its politics: at least Robert Owen is written off as a 'political quack', and the chances of the unmarried young members of an Owenite community remaining chaste derided—just as the Godwinian sexual utopia was derided by Malthusians. The coarseness of these attacks on socialism, with their jokes about socialist 'erections' and socialist lecturers called 'Slowcock' and 'Bullcock', creates just the bemusing mixture of ideological statement and smut which is so characteristic of the magazine. Incidentally, the *Town* also carried advertisements for some of the anti-male midwife literature, which harped incessantly on the impossibility of male doctors subduing their sexual drives in a clinical setting.[96]

The *Town* attacked both freethinkers and puritanical Evangelicals. It was bound to turn on the latter in self-defence, but it does sound genuinely affronted and alarmed by the Evangelical project of prostitute rescue work being performed by respectable young women. The converse of this is that key element in the classic moralist position on transgressive sex, namely, approval for officially regulated prostitution. Nicholson ran an extraordinarily long series of over a dozen editorial pieces on this question.[97] The principle of tolerated but controlled pre- and extra-marital sex, offered by officially licensed prostitutes, would tend to reconcile the paradoxes in the *Town*'s handling of sexuality, and in Nicholson's career in general: the frank titillation and advice offered to the man-about-town, coexisting with a claimed distaste for transgressors such as paedophiles and adulterers, and concern for young women's purity. Even the *Town*'s oddly vehement attacks on brothel-keepers acquire some logic.

---

[95] Nicholson 1965: 218; *Frasers* 17 (1838), 288–90.   [96] *Town*, 1124, 1158, 1166, 1204.
[97] *Town*, 155, 604, 612, 620, 628, 636, 644, 652, 660, 668, 676, 685, 692, 700.

It is in fact by no means inconceivable that Renton Nicholson, a major sponsor of dirty-mindedness and the fast life in Victorian London, was a perfectly honest editor of the seriously Whiggish *Crown* (though an attack on the *Town* in the first number of that magazine makes a tongue-in-cheek attitude more probable).[98] In the great division of opinion on prostitution policy, which I shall seek to sketch shortly, mainstream Anglican thinking for many years came down, if not in favour of licensing, then at least against the prostitute-rescue approach. In my view classic moralism—the age-old anti-sensual, but anti-idealistic, secular code—contributed much to the formation of this Anglican response to the 'great social evil'.

And this code was one which through its non-Christian antecedents in the classical tradition, its distrust of utopianism, and its liking for the appeal to supposedly fixed facts about human nature and society, always tended to leave the door open to a degree of acceptance, however sullen, of sexual transgression. I have tried to trace this logic in the scurrilousness of the definitely moralistic *Scourge*, and of the less definitely moralistic *Town*. This has yielded the strange outcome that the great Anglican apologist William Paley, for all his brisk censoriousness about non-marital sex, belongs in the same camp as the notorious Renton Nicholson. At best, this is only a partial truth, but that it contains at least some truth may be confirmed by reference to the sexual attitudes of the dominant weekly periodical of the Victorian era, the *Saturday Review*.

The *Saturday Review* was founded in 1855. At the time the two best-selling high-brow weeklies were the *Spectator* and the *Athenaeum*. The *Saturday Review* had overtaken the first by the end of the 1850s, and the second by the end of the 1860s. It was recognized to be the main organ of educated opinion in the third quarter of the century and also to be aggressive, destructive, and even flippant in its essential tone. It acquired nicknames such as The Saturnine, and The Saturday Slasher/Snarler/Reviler.[99] Every historian of the period knows of the importance of the *Saturday Review*, and yet it has been the specific subject of only one modern study, a book published in 1941. Its ascendancy in the Victorian world has never really been addressed in modern accounts of the Victorian mind and temperament; it is not even cited by historians as much as its contemporary success would seem to warrant.

[98] *Crown* (1 July 1838), 4, and (29 July 1838), 5.   [99] Bevington 1941: 22–4, 43.

One reason is that the *Saturday Review* itself tends to force on the historian a choice between Victorian traditions. It disliked, attacked, and tried to set a different rhetorical and mental example to, commentators such as Carlyle, Ruskin, and Arnold. Our age has been much more drawn to these writers. We have difficulty with them, but we seem to like them in some sense we scarcely recognize, because we have not been diverted from regarding them as the great voices of the era, even though the *Saturday Review*, with an impressively large readership behind it, insists that they are not.

In politics the *Saturday Review* was conservative. In religion it was the essence of latitudinarian Anglicanism, opposed to the low church and dissent, but also (despite the loyalties of its founder, Alexander Beresford Hope) to Anglo-Catholicism. Not that the *Saturday Review* wished to utter any doctrinal claims on behalf of mainstream Anglicanism; this was regarded as a valuable religious force precisely because it made only hazy dogmatic demands. On moral and social questions the *Saturday Review* had what its historian calls 'a fundamental quarrel . . . with Victorian manners and morals' because it hated the sentimentality of the age: 'it may reasonably be doubted whether there has ever been a more thoroughly sentimental time than the present'.[100] Our first reaction is to agree; our second reaction should be to wonder if Victorian England really can have been as 'thoroughly sentimental' as we suppose, if that tendency was consistently deplored at the time by so powerful an organ of opinion.

The 'sentimental' in morality denoted for the *Saturday Review* both leniency and an artificial or idealistic inflexibility of standards. The converse position is the typical classic moralist mixture of severity and respect for supposed realities. Thus, the *Saturday Review* was keen to urge that morality 'does and must vary slightly from time to time'—while it was also in favour of flogging and public executions. It greatly disliked the contemporary Evangelical-minded advice literature for girls, with its urgings to an improbable regime of good works, serious reading, and self-denial. But the organ which published Eliza Lynn Linton's famous 'Girl of the Period' article was also predictably severe on the least forwardness and flirtatiousness in women's behaviour and dress, leaving its female readers a narrow path of Joan Hunter-Dunneish sexless affability to tread.

When women did transgress the Evangelicals were again wrong,

---

[100] Bevington 1941: 98–9.

according to the *Saturday Review*, in supposing that they had been the unsuspecting victims of male ploys: on the contrary, 'very few men in good circumstances have the inclination, and fewer still have the opportunity, of looking out for beauteous virtue in order to corrupt it. And, to speak the plain truth, even those who . . . might be disposed . . . are often saved their trouble by the advances which beauty . . . is disposed to make to them.' And anyway, for good measure, such women had been morally undermined by the 'rigorous Puritanism' of 'a certain party of the clergy', who had made casual sex the only recreation available in modern cities. The *Saturday Review*'s most prolific contributor, Fitzjames Stephen, also penned in the *Cornhill* one of the period's strongest defences of the double standard.[101] The fully fallen woman, the prostitute, should be treated like the cross between the criminal and the down-and-out which she is: she 'has just the same claim upon society . . . as any other member of the dangerous classes; and . . . is as much entitled to charitable succour as any other person engaged in a life immoral, disgusting, and ruinous both to body and soul'. This formula was designed to make Evangelical rescue work seem hopelessly lenient on prostitutes— even guilty of a degree of 'subtle indelicacy and . . . refined coarseness' itself. But the *Saturday Review* was by no means opposed to letting these dangerous and disgusting women ply their trade if they wished: in fact it was 'Puritanical prudery' when church-based groups sought to strengthen police powers against prostitutes, or opposed regulation and licensing along European lines.[102]

The official registration of prostitutes was already a reality in Britain when the *Saturday Review* was uttering some of these remarks, though the fact was not generally noticed. In 1864 the first of three measures collectively known as the Contagious Diseases Acts was passed, giving powers to local police and JPs to order the medical examination of women identified as prostitutes, and to confine them in certified hospitals if they were found to have venereal disease. The successive acts all applied to the neighbourhood of military garrisons and naval dockyards: the specified areas were enlarged, and the list added to, but proposals for an extension of the system to purely civilian areas came to naught.

---

[101] Ibid. 99–103; *Saturday Review*, 5 (1858), 239, 11 (1861), 336–7, 16 (1863), 460–1, 19 (1865), 371–2, 20 (1865), 396–7, 22 (1866), 5481–2; 'Anti-Respectability', *Cornhill*, 8 (1863), 282–94.

[102] *Saturday Review*, 5 (1858), 84, 10 (1860), 417–18, 21 (1866), 748–9.

The rationale for this legislation was always medical, though its goal—of prostitution both stigmatized and permitted as a practical necessity—is the one dear to classic moralism. Medicine in the Victorian period, as a source of secular and empirical authority for anti-sensual views, tends to form a serious wing of classic moralism, more solemn than the *Saturday Review*. Peter Gaskell, in his furious attacks on the sexually depraving effects of the factory system, is a notable early example of a doctor taking a severe and gloomy line on contemporary sexuality from entirely secular and scientific premisses. The *Lancet*, habitually, has a high moral tone on sexual questions, but without any appeal to religion. Its anti-sensualism is not directly empirical, but nor can it be accused of a dependence on any author-ity outside the medical—so there is a general if vague sense of moral-ism validated by medical wisdom, and this despite the radical politics of Thomas Wakley.

William Acton, one of the architects of the regulation of prostitu-tion, is bemusing on sexual-moral questions in the way that classic moralism often is. His celebrated *Prostitution*, first published in 1857, combines tolerant and even affirmative feelings about prostitutes and the causes of prostitution with calls for a 'crusade against vice', in which it is even suggested that the example of the Roman Catholic Church, rather than that of the Church of England, should be fol-lowed. There is tolerance and relish for London's night-life in the mere scale and detail of Acton's treatment of the subject, while pros-titution is consistently represented as an interlude in the life of healthy and good-looking girls—who are accordingly attractive, well-groomed figures on London's pavements. The male sex-drive is the great engine of prostitution and can scarcely be overrated: the prosti-tute rescue movement is an absurd futility. But Acton also urges that a suitable moral and physical training, including gymnastics and cricket, can transform contemporary sexuality. As I have explained in the first part of this study, this is the theme of his *Functions and Disorders* writ small. In that context it has the appearance of being the medical equivalent of ideas on the plasticity of the libido voiced by political progressives. In *Prostitution* it is puzzlingly conjoined with policies and attitudes apparently favourable to the routine pre-marital satisfaction of male sexual needs.

Acton has been forced into incoherence by trying to fight quixotic religious moralism on one front and sexual freedoms on another. Varnished over the whole, and giving a certain aspect of cogency, at least by classical moralist lights, is a persistent appeal to the empirical. According to Acton, it is a fact to be faced that the human soul is depraved—which is where Roman Catholicism is more wise than Protestantism; the 'moral and religious curb' is ineffectual. But wisdom in a more scientific vein teaches that the human mind is far from irretrievable: much can be achieved by a 'worldly-wise and physico-moral training', an 'anti-sensual instruction'. It is a pernicious deduction from the damage caused by the more traditional kinds of restraint to argue that no restraint is possible or desirable: desire for sex as a purely physical experience is 'unnatural' and if indulged would lead to 'a demoralized condition of mind and body'. The conjunction of the 'worldly-wise' (conceived in antithesis to religion) and the 'anti-sensual' nicely conveys the mentality I have been analysing as 'classic moralism'. With the addition of the 'physico-moral' this traditional code takes on a very Victorian colouring.[103]

An earlier commentator on prostitution whom Acton greatly admired is W. R. Greg, and Greg represents perhaps the most intelligent attempt by classic moralism to weld its various impulses into a cogent statement on this topic. Greg is right at the edge of classic moralism, as far from the centre as one could go in this baggy school of thought. His article on prostitution actually appeared in the *Westminster Review*, organ of utilitarianism. Greg had received a Unitarian education, and became an avowed disbeliever in orthodox Christianity; a year after the prostitution essay he published a book-length critique of the veracity of the Bible (though not, as he put it, of 'the Faith of Christ').[104] But he was in reality a political reactionary, and his views on prostitution are quite different from those of real *Westminster Review* men, such as John Chapman and Mill. He locates himself in the prostitution debate very much where *Saturday Review* readers would probably have located themselves. He is 'lax' by the standards of religious morality, he says, but 'savage and absurd' by the standards of 'the mass who follow current ideas without thinking at all'. His morality on extra-marital sex is a graded and relative one, but severe enough: 'sexual indulgence . . . is, when accompanied by love, a sin *according to nature*; fornication is a sin *against nature*'.

Greg does, in his leniency towards the prostitute, move closer than the *Saturday Review* would have dreamt of doing both to the

---

[103] Acton 1857b: 161, 163–6, 173, 1870: 114–16.   [104] Greg 1850, 1851: p. ix.

magdalenism of the Evangelicals, and to the progressives' belief in
male and female powers of continence. 'Every heart should bleed',
he says, 'for the position of an English prostitute, as it never bled at
any form of woe before'. Not only is lust never the force which drives
a woman to fall at first, but female desire does not exist until intercourse has been experienced. Both the respectably married and the
seduced feel a disgust for sex at the beginning, according to Greg:
the sense of shame is the same in both cases. But on male desire,
Greg shows how his loyalties are ultimately with classic moralism.
He hints that his true belief is that men, though they do feel desire
before the experience of intercourse, can subdue this until marriage,
however late. He will refrain from stating as much, however, because
it sounds 'puritanic or Quixotic', 'romantic or high-flown'. Even
Greg's magdalenism yields, in the end, to regulation, which he supports: indeed the 'woe' of the English prostitute's life, he suggests,
would be much abated if these women had a legally recognized role.

Chapter Two

༺ ༒ ༒ ༒ ༒ ༒ ༺

EVANGELICALS AND SEX

༺ ༒ ༒ ༒ ༒ ༒ ༺

I

Acting against Vice

THE spectacle of a man like W. R. Greg, at the mid-century, speaking of the 'sin' of all extra-marital sex, and implying the possibility of complete chastity for all men before marriage, even though he is ostensibly opposed to the 'puritanic', may make one feel that the 'puritanic' actually has him well and truly in its grip. The date is certainly late enough for the hypothesis I mentioned earlier, namely, that Victorian anti-sensualism is chiefly a matter of Evangelical moral attitudes, shorn of their doctrinal setting, finding their way into the thinking of men and women situated well beyond the boundaries of formal Evangelicalism. By 1850 there was plenty of time for beliefs articulated at the end of the previous century to have carried far and wide. This would obviously amount to a very different kind of explanation from the one I have offered, in terms of *Saturday Review* classic moralism, for the attitudes of a Greg or an Acton. And the hypothesis could also apply—indeed *is* very commonly applied, at least by implication—to the anti-sensualism of men like Mill and Spencer, which I see as progressive anti-sensualism, with its roots in the utopianism of the 1790s.

We are concerned here with Anglican Evangelicalism and its impact on educated opinion, that is, with the influence at the middle- and upper-class level of William Wilberforce and Hannah More, of the Cambridge dons Isaac Milner and Charles Simeon and their followers, and of the 'Clapham Sect' of wealthy London professional

and commercial men. Wilberforce was converted to vital, zealous
Christianity in 1785. As I noted in the last chapter, the Evangelical
element in the Anglican ministry grew steadily, even for a time
explosively, from about this date, down to the mid-nineteenth century. Against the possibly less impressive scale of conscious allegiance to Evangelical principles among lay people must be set the
huge breadth and energy of Evangelical proselytizing work and the
atmosphere this created. But the full human reality of Evangelicalism's influence is hard to assess. Indeed, for a phenomenon
referred to so glibly in the history books, it is very poorly understood.
One of its recent historians has conceded that 'our ignorance of
Victorian Evangelicalism is profound'.[1]

I shall concentrate on certain doctrinal questions. In any picture of
the period which gives a central role to Evangelicalism it is agreed
that, somewhere along the line, the Evangelical creed was largely
stripped of its specifically religious components as it moved into an
ascendant position in people's moral codes. Victorian men and
women, it is claimed, embraced the principles of sobriety in dress
and speech, strict observance of the Sabbath, close moral control of
household activities and children's education, ostracizing of the sexually lax, and so on, while they believed only partly or not at all in the
key theological doctrines of Evangelicalism: the absolute requirement of sustained Christian conversion for salvation, their own
achievement of the latter (together with a very small fraction of the
human race), the literal truth of scripture, and the direct operation of
God's providence in everyday life.

It is often remarked how many illustrious Victorians came from
Evangelical backgrounds but themselves foreswore Evangelicalism—
with the implication that here Evangelical morality, but not
Evangelical metaphysics, stuck and then transferred itself to the
larger culture. Cases in point are Macaulay, De Quincey, the
Brontës, George Eliot, Ruskin, Peel, Newman, Jowett, Kingsley, and
Manning. But it is fair to put the emphasis, for a moment, on the element of rejection of parental hopes and teaching in these careers, as
opposed to the element of inheritance. 'Manifestly all was not well
with the party':[2] even Wilberforce's own sons did not accept their
father's creed. Perhaps some of the problem did lie in moral matters.
If it is accepted that the Evangelical legacy could be unstitched, and

---

[1] Englander 1988.   [2] Newsome 1966: 9.

only accepted in part by its heirs, it is important to be clear what that legacy looked like—and, in particular, what it presented in the way of teaching on sexuality.

One line of thought on this whole question distinguishes, in effect, between two waves of English Evangelical activity, and explains the supposedly split reaction of Victorians to Evangelicalism—overt rejection but tacit reception—in terms of these different phases. In Ford K. Brown's enthralling, idiosyncratic, and astoundingly ambivalent study (the encounter with which is one of the pleasures awaiting the enquirer into this subject) the distinction between the Evangelicalism of Wilberforce's generation and that of the early Victorian years is frequently drawn. It is in the second wave, according to Brown, that Anglican Evangelicalism takes on its repellent physiognomy: the 'rigid, hard, mannerless pedantry', the 'puritanical strictness and senseless rigour . . . hardening doctrinaire convictions defying humaneness and good thinking'. He suggests that the famous caricatures of Evangelical types in Victorian fiction—by Dickens, Charlotte Brontë, Mrs Trollope, and Wilkie Collins, for instance—generally have the second generation in their sights. A typical representative of this generation was the Reverend William Carus Wilson, the model for Mr Brocklehurst, proprietor of Lowood School in *Jane Eyre*. Brown does not believe that the portrait is overdrawn.[3]

Charlotte Brontë was at school in the mid-1820s. Did she, or at least a significant number of her contemporaries, though alienated by the extremes which Anglican Evangelicalism had come to by this date, also imbibe the spirit of the Evangelical revival as launched by Wilberforce forty years earlier? Something of the sort seems to be a widespread working assumption with students of the period, but it is very seldom exposed to discussion.[4] The keynote of the first generation's demeanour had been (to use Brown's words again) 'playfulness, jocularity, mirthfulness . . . "apostolic joy" . . . a proper suavity and elegance of manner'. It was a style designed to allay Anglicans' anxieties about what was being asked of them, to suggest 'a happy religion not unmethodistically unsuited to the great'. It had its undoubted success. One commentator perceived the fact that 'the profession of religion is not only tolerated, but, to a certain extent, has become fashionable'—as though polite society had converted the Evangelicals, rather than vice versa: 'The progress of general educa-

<hr>

[3] F. K. Brown 1961: 445–57, 506, 508.   [4] See, however, C. Hall 1979.

tion and refinement has considerably modified the sentiments, habits, and taste of the professors of religion . . . the harsh features of puritanical piety . . . have been softened down; its contracted views and illiberal feelings exchanged for opposite qualities.' Even a writer hostile to the moralism of his day, Richard Shannon in the *Edinburgh Review*, sees it as the context and climate out of which Evangelicalism was then growing (in 1831), and not as in any way the work of a better camouflaged, earlier contingent of Evangelicals.[5]

There was no doubt a good deal of reflex antagonism to Evangelical moralism which, if anything, meant the opposite of what it said: as Augusta Webster observed about Mrs Grundy, 'each of us fires off his regulation volley of heresies and epigrams without any real intention of damaging her . . . Our very reviling is a homage'. When Lucy Aikin said in 1828 that 'the doctrines called Evangelical make all the noise' but 'their success is at its furthest' she was perhaps whistling against the wind.[6] The two-generations model clearly has some relevance. However, it does not help with the fact, or alleged fact, that Victorians swallowed Evangelical moralism, specifically, but gagged on Evangelical theology. And, more importantly, it requires us to overlook the very considerable scale of the conscious and articulated hostility to Wilberforce and his colleagues, that is, to the first phase of the revival.

For all its camouflage of 'suavity' and 'jocularity' Wilberforce's campaign never completely succeeded in driving a wedge in the public mind between the new Anglican Evangelicalism and less congenial social and religious groupings. For some commentators these linkages placed the new movement quite beyond the pale. As Ford K. Brown has put it memorably, 'it is a chastening thought that religious reformers so devoted to their country's good, so desperately concerned . . . not to be taken for Methodists, dissenters, political radicals, liberals, or any kind of people disliked by the great, should have been violently charged with just such corruptions through the most crucial decades of their campaign.' The great vehicle of these charges was William Gifford's *Anti-Jacobin Review*. It did not avail that, politically, Wilberforce and his parliamentary associates supported anti-popular and anti-democratic policies with complete persistence: voting for all Pitt's oppressive Acts of the 1790s and the post-Napoleonic period, and defending the army's action at Peterloo.

[5] F. K. Brown 1961: 456; Thomason 1827: 14–15; Shannon 1831.
[6] Webster 1879: 32–6; Channing 1874: 29.

Nor that Evangelicalism through its arm in the Proclamation Society and the Vice Society prosecuted radical publications by Paine and others. (All this was well-known to Francis Place, who called Wilberforce a 'devil'.)

With more justification, the *Anti-Jacobin* could not be thrown off the trail of associating Wilberforce's movement with 'a morose and harmful puritanism'. In practice Evangelicals had co-operated with dissenters in many of the missionary and reforming bodies, and this tainted them in the eyes of their enemies with a deeper theological association than they would ever have wished to claim. As I have just mentioned, the use of the term 'methodism' to cover the Wesleyan churches and the Anglican revival, if perhaps always somewhat hostile, is quite standard.[7]

What did first-generation Anglican Evangelicalism stand for as far as sexuality was concerned? There was, as emerges in the controversy over Malthus, a tendency in Evangelical and dissenting circles to elevate marriage, and the sexual impulses which impelled young people into marriage, as aspects of God's providence which it was wrong to resist. The celebrated Evangelical bowdlerizer James Plumptre argued that celibacy actually interfered with the spiritual life: 'self-denial . . . may be carried too far. The body is to be preserved in health and vigour, that the soul may be enabled to discharge its functions.'[8] But the realm of transgressive sexuality started as soon as the tightly drawn frontiers of marital sexuality had been crossed, and right through this wide territory Evangelicals appear to have been as consistently anti-sensual as was humanly possible. The Evangelical restraints on dancing, novel-reading and playgoing, ornamental or revealing dress for women, and so forth, one expects to be accompanied by a very severe code on sexuality in language, on eroticism in the relations of the unmarried, on any immoderacy in marital sexuality, and on children's exposure to sex—and there is every indication that this was the case, and honestly the case.

There are theological routes by which pro-sensualism, in some departments of life at least, can be reconciled with anti-sensual strains of Christianity. I have traced a few examples in section ii of Chapter 1. But with the arguable exception of the magdalenist policy on prostitution, pioneered by certain Evangelicals, Anglican

---

[7] F. K. Brown 1961: 113, 162, 174, 369, 503.
[8] Plumptre 1812: 12.

Evangelicalism does not seem to have been lured down these routes, for good purposes or bad. The Wilberforcean affability, elegance of life, and 'joy', which may have nudged Henry James Prince in the direction of a startling sensualism, was never a morally undermining code for Wilberforce himself—even though he seems, for example, to have eaten well and drunk alcohol in fair measure. Much later the 'Recordite' party is accused of being drawn close to moral licence generally by its antinomianism but, it is admitted, not over the brink: 'the very preachers who proclaim the "imputed righteousness" of the most sinful believer, seldom proceed to the conclusion of the Antinomian,—"Let us continue in sin that grace may abound"'.[9]

Ford K. Brown notes that 'only one charge of "personal immorality" (in a vulgar physical sense)' was made against an Evangelical clergyman in all the onslaught on Wilberforce's movement. To this may be added some talk of flirtatiousness in the behaviour of popular Recordite preachers, but there was a vanishingly small body of accusations in this vein compared with the habitual smearing of Nonconformist divines.[10] Nor was there anything comparable to the propaganda against the sexuality of Methodist devotional imagery and behaviour. To the world at large (apart from accusations about the indelicacy of female involvement in the rescue of prostitutes, which I shall come to later) the only thing possibly wrong with Evangelical ideas about sex was their severity. Puritanism and prudishness were the points for attack. The obvious suggestion was made, as by the *Scourge*, that to see a dirty possibility was to betray a dirty mind, but this was said more to score a small debating triumph than as a serious accusation of immorality.

It is quite possible that Evangelicalism, through the weight and persistence of its activities on behalf of a stricter sexual standard, managed to mitigate the effects of popular resistance to the puritanical and the prudish—and achieve a conversion of the nation to antisensualism in some sense in spite of itself. The tally of overt Evangelical effort in this direction does not make this seem very likely, however. Evangelicals certainly wanted to have an impact on sexual codes, and sometimes put considerable effort into the project, but the net result does not look like the springboard for Victorian anti-sensualism.

Wilberforce's own activity on behalf of sexual controls was inter-

---

[9] Conybeare 1853.   [10] F. K. Brown 1961: 402; Conybeare 1853.

mittent, if notably committed. There can be no mistaking his abhor-
rence of sexual laxity. The very first of the scores of national societies
for moral and religious purposes founded by Evangelicals was the
Society for Giving Effect to His Majesty's Proclamation against Vice
and Immorality. Wilberforce brought it into being in 1787 in
conscious imitation of seventeenth- and early eighteenth-century
reforming societies, having already induced the Privy Council to
issue the royal Proclamation in question in June of that year. This
scatter-shot exhortation to citizens, legal officers, and clergy to come
down hard on all 'vice, profaneness and immorality' was very widely
circulated, and the society sought for the next dozen years or so to
maintain private and public vigilance on the legal aspects of morality.
It also mounted several prosecutions itself, including some for
obscene publication.

The committee of the Proclamation Society was not distinctively
Evangelical, and the motives of many of its members must be
accounted those of classic moralism. From 1802, however,
Wilberforce and several other leading Evangelicals took an increasing interest in the running of a body which had probably been
founded the year before simply as an auxiliary to the Proclamation
Society, known as the Society for the Suppression of Vice and the
Encouragement of Religion and Virtue. This then merged with and
effectively eclipsed the old Proclamation Society in 1804. The Vice
Society thereafter was more continuously manned by the Evangelical
élite, Wilberforce included, than any other body with which they
were involved: Shaftesbury was a vice-president in the mid-1870s.[11]
Nevertheless this side of the Evangelical operation was notably
unprosperous, probably more so than anything else it attempted.
There were practical difficulties in acting against 'vice, profaneness,
and immorality' in the way both societies tried to act, but the main
reason for their failure was that the project was out of tune with the
culture.

The Vice Society is admittedly an obscure organization, and even
the broad outlines of its history are not clear. The transition from the
Proclamation Society is not well understood, in terms of timing and
personnel. Roughly speaking, an Evangelical-dominated Vice Society
replaced the Proclamation Society in 1804, but the latter may have
had some kind of continuing existence. In 1804 the Vice Society was

[11] Radzinowicz 1956: 149–65; Bristow 1977: 38–41; F. K. Brown 1961: 83–8, 428–9;
National Reformer, 30 (1877), 529–30.

buoyant, with a membership probably of at least 1,000—presumably boosted by whatever merger had taken place. By the 1820s this halcyon total had been cut by no less than three-quarters. The membership was probably now concentrated in London, for it is only in the early years that there is any record of subsidiary societies in the provinces. But at a modest level of activity the society did prove to be a survivor, and made appearances in the courts down to 1880, perhaps experiencing something of a revival of energies towards the latter date. This period of more visible 'attitudinizing as purifiers of society' was short-lived, spoilt partly by a scandal concerning misappropriation of funds. In 1885 the National Vigilance Association was founded; the Vice Society made a financial contribution, but must soon after have been eclipsed by that highly successful organization.[12]

Contemporaries found the Vice Society mysterious. Its failure to publish reports and membership lists was often criticized. As early as the 1830s Michael Ryan was asking 'what has become of the Society for the Suppression of Vice?' and some years later it seems to have been confused in the popular mind with prostitute rescue societies. Its very existence could be unaffectedly doubted by the 1850s, and in 1869 the *Quarterly Review* called for the setting-up of a moral reform society on late seventeenth-century lines, as if such a revival had never been attempted.[13]

The society's invisibility, for most of its life at least, was the joint effect of the nature of its work and its unpopularity. It was unique among all Evangelically derived bodies in tackling moral questions by means of the law. The entire principle of its operations was the bringing of the powers of the police and the courts to bear on moral transgression (with the creation of the right legal framework sometimes a prior necessity which the society lobbied for: as in the case of well-known Campbell's Act of 1857 against pornography). This was the principle founded, or revived, by Wilberforce with the Proclamation Society—which not only exhorted the agents of the law to act against immorality but elaborately specified the statutes avail-

[12] *London Gazette* (9 June 1860), 339–42; Howse 1953: 120; Society 1803: part ii, p. 95; Bristow 1977: 42, 48–9; F. K. Brown 1961: 429; *Hansard*, NS 8 (1823), cols. 709–35; Quinlan 1941: 205; S. Smith 1859, 132; Radzinowicz 1956: 156; *National Reformer*, 29 (1877), 402; Conway 1878: 11; Prochaska 1980: 214.

[13] Ryan 1837: 15; *The Times* (16 Oct. 1845), 8 (case of theft from A. Gasby); Ritchie 1857: 22; *Quarterly Review*, 126 (1869), 448–79.

able to them. As an MP Wilberforce several times supported measures which bolstered the penal system.[14]

Bringing, or assisting, successful prosecutions is more difficult and more expensive than pressing a few thousand tracts on passive working-class people. Yet the Vice Society, like all its peers, was supported by private donations only. It is known often to have been badly short of funds, not just in its later years but as early as 1810. It did not bring nearly as many prosecutions as it would have wished. But there was a sense in which the society did not need public assistance. What advanced its work most were skills and power in the criminal domain. With a receptive magistracy and a like-minded police force its aims became much more achievable. Both before and after the creation of the metropolitan police in 1829 there was co-operation in a varying degree between the society and London's police. The society occasionally fell back on the expensive recourse of paid private agents.

What is more, its leaders and supporters also had a lively sense of the hostility felt towards them in some quarters: there was not, they felt, 'so unpopular a Society in the kingdom'. One receives the impression of the Vice Society for most of its long life, once the cheering of the first few years had died down, withdrawing into cautious and deliberately unobtrusive activity in its single London base in Lincoln's Inn Fields, under the guidance of its extraordinarily long-serving secretary George Pritchard (or Prichard). The careful, inconspicuous work continued to achieve occasional highly visible results, such as Campbell's Act and the very unpopular 1878 prosecution of Edward Truelove for publishing birth-control literature. (The even more celebrated prosecution of Charles Bradlaugh and Annie Besant, the year before, for the same offence, seems not to have been due to the Vice Society, despite much speculation to this effect: it is a token of the unpopularity perceived to attach to such attempts that the organization who funded this action has never been identified.) But when there was a flurry of lay and clerical agitation over London prostitution in January of 1858, for example, it seems, to judge from *The Times* reports, that meetings on the question held in the society's premises were not as well-attended or significant as those taking place elsewhere.[15]

---

[14] Radzinowicz 1956: 159, 489, 498–504.

[15] R. Knox 1857: 189; Society 1810, 1868; *Lancet*, 1873(1), 250–1; *Hansard*, 46 (1857), col. 1355; Bristow 1977: 45–7; Radzinowicz 1956: 175–6; PP 1817, vii. 391; *Christian Review* 2 (1828), 10; *The Times* (4 Jan. 1858), 4, (7 Jan. 1858), 10, (16 Jan. 1870), 9, (1 Feb. 1870), 9.

Interestingly, the society's cause—in so far as this was sexual immorality, under some definition—was not in itself the reason it was disliked. Not only the confused voice of classic moralism in the *Scourge*, as we have seen, but also men of the left—Francis Place, Richard Carlile, and Joseph Hume—picked out the society's campaign against indecency as the valuable part of its operation.[16] When the National Vigilance Association sought support for a campaign of legal measures against sexual immorality very similar to the society's it filled not only its coffers but a large area of Hyde Park. At about the same time the Vice Society was calling it a day. It is arguable, in fact, that in the decades running from about 1820 the Vice Society represented a traditional moralism which was out of tune with the new moralism shared, paradoxically, by the new Evangelicalism and secular progressives—which was at once more stringent and more high-minded on sexual questions.

To start with, the society's treatment of sexual morality was tinged with the expediency more commonly associated with classic moralism, very surprising to encounter in Wilberforce's moral environment, but perhaps due to his sense of what was tactically required. The society's first statement to the public is dated 1803, when there was apparently as yet no direct link between the Evangelical leadership and the society, and was drafted by a high church writer, John Bowles. It is pointedly unfanatical on sexual vice: the 'complexion of the times' tolerates many things, 'indecency at one period of civilisation, is considered as decency at another' (later the thinking of the *Saturday Review*), and nothing will be achieved without 'moderation' on these questions.

This remained part of the society's moral legacy, even after the Evangelicals had had a chance to purge its doctrine. Pritchard, giving evidence to the Commons Select Committee on the Police of the Metropolis in 1817, was not disposed to complain about the lack of legal machinery available to the society for the suppression of prostitution, because 'there are certain evils which it seems necessary to tolerate in order to prevent the greater; and the Society . . . conceive that the only useful and practical point to be aimed at is the mitigation of these nuisances'.[17] One wonders what Wilberforce made of such remarks. He might have regretted them as a necessary concession to worldly men. But perhaps he would have concurred with

---

[16] Quinlan 1941: 221.

[17] Society 1803: 62–3; PP 1817, vii. 534. Though see also Radzinowicz 1956: 192–3.

their tacit relinquishing of the possibility of doing anything about a prostitute's soul, and confining of the task of Evangelicalism, at this level, to a cosmetic exercise in street decorum. If so, he was notably out of step with many Evangelicals of the next generation.

Secondly, the Vice Society as established right at the outset of the nineteenth century had many targets in addition to sexual indecency—Sabbath-breaking, very notably, but also blasphemous publications, lotteries, cruelty to animals, and even weights-and-measures frauds—which were not necessarily regarded as less reprehensible in the Victorian period, but were seen as less pressing or less amenable to a punitive solution. It is not easy to judge the true priorities of the society in relation to these various kinds of abuse at different periods. The parent body, the Proclamation Society, brought a few obscenity prosecutions, and it has been claimed that this kind of work loomed even larger in the efforts of the Vice Society.

This may have been the case for a time, but the record of known prosecutions shows that Sabbath-breaking, and even lotteries, were for many years still targeted much more than sexual immorality of all kinds (prostitution, brothel-keeping, and sale of obscene materials). Indeed, in the sum of all Vice Society prosecutions down to the 1850s the number of cases in the latter category must have been in the low hundreds, at the most, but the number of cases brought for violation of the Sabbath ten times that number. At some point after the mid-century obscene publications and pictures do seem to have become the society's main preoccupation, with an accelerating activity against these right down to 1880.[18]

There were occasional parliamentary initiatives in the period on aspects of personal sexual morality, falling under two heads: adultery and seduction (the latter theme evolving from parliamentary concern about prostitution). Evangelicals were predictably a force on the side of severity in these proceedings, though it could not be said that they played a rawly prominent role, and much of the impetus for the initiatives came from classic, non-Evangelical moralism. In 1800, not for the first time, an attempt was made to forbid the marriage of the adulterous parties in a divorce suit. A bill was introduced in the upper house by Lord Auckland, who was not connected with the Evangelicals or even with the Proclamation Society. Its most

---

[18] Bristow 1977: 39–42, 49; Quinlan 1941: 218–19; Radzinowicz 1956: 171–2; Society 1803: part ii, pp. 87–91; Society 1812, 1868; Ryan 1839: 111–16; *Hansard*, NS 8 (1823), cols. 709–35, 146 (1857), col. 1355.

vigorous supporter in this chamber was the ageing Samuel Horsley, Bishop of Rochester, who had been a founding member of the Proclamation Society. It was also spoken for, less strikingly, by Beilby Porteus, the moralistic Bishop of London, not a member of Wilberforce's circle either, but a notable friend of Evangelicalism and probably the most vigorous president of the Proclamation Society.

The bill was approved by the Lords but thrown out by a sizeable majority in the Commons, where it had acquired an amendment making adultery a misdemeanour in law (Auckland would have liked substantial prison sentences for the offence). Wilberforce supported it in a speech which made very strong claims for the centrality of sexual probity in Christian doctrine. According to his exposition, Christ had not only denounced adultery, but had 'dwelt upon it more particularly' than any other sin; it was a national problem 'of much more importance than any question about peace or war, or any constitutional question'. And Wilberforce drew an important distinction between 'the moral decency so characteristic of these times', which was just a matter of 'decorum of deportment', and the reality of contemporary religious and spiritual corruption, as signified by adultery. As well as being an interesting piece of evidence for a new respectability in English manners which had nothing to do with Evangelicalism (and in fact is here disavowed by the movement's leader), this touches on an important and ambiguous theme in the Evangelical attitude to sexuality.

Wilberforce still contemplated trying to have adultery made a criminal offence, and John Bowdler seems to have wished the death penalty for it, but severe punishments for adultery and seduction were also favoured by moral alarmists of a non-Evangelical cast, such as Patrick Colquhuon and the Bishop of Chichester. Auckland's own disquiet on adultery was so strong that he returned to the Lords in 1809 to propose a standing order requiring all individual bills for divorce to have attached to them a prohibition against the marriage of the parties specific to that suit. The order apparently came in, despite objections in the Commons—but was presumably discontinued at some later date.[19]

The mercenary sexual corrupting of young women—sometimes referred to as 'seduction' but less confusingly thought of as procure-

[19] *Hansard*, 35 (1800–1), cols. 320–1; Jaeger 1956: 44; Radzinowicz 1956: 193, 201, 254–5; Horstman 1985: 15.

ment—emerged as the solitary viable principle in various attempts to legislate against prostitution that were made in the period 1844–9. This emphasis partly reflected the prejudices of the sponsors of the legislation, but also came about because other lines of statutory attack on prostitution were foiled by special interests. The result is that the one statute enacted—the very brief Protection of Women Act of 1849—has a 'magdalenist' spirit (in a sense I define in the next section) which is to some extent accidental. In the creation of this magdalenist Act there was no particularly noticeable contribution, whether outside parliament or inside, from Evangelical opinion. Certain religious figures played a part, but they were middle-of-the-road, conservative churchmen.

In 1844 Henry Philpotts, Bishop of Exeter and the most notable conservative in the Anglican élite, introduced in the Lords a bill which sought comprehensively to undermine organized prostitution. It had penalties and prohibitions not only against brothel-keepers and their staff, and those who colluded with them in debauching girls, but also against the owners of such premises if their property reverted to use as a brothel once a conviction had been achieved. As amended in committee, the range of measures likely to hurt the brothel-industry was reduced (in particular, there was now no reference to the owners of brothel-premises), while procurement was made a specific separate offence, of comparable weight to running a brothel.

In this version the bill was still consonant with the basic philosophy of the body which had persuaded Philpotts to sponsor it, namely, the Society for the Protection of Young Women (as this organization expanded its operations beyond London it took the name 'Associate Institute'). Even as amended, there was still a striking avoidance of any criminalizing of the prostitute: in the bishop's words to the House, 'there was nothing . . . that had any relation whatever to the suppression of prostitution'. But while the Associate Institute evidently wished, otherwise, to work on a broad front for the defence of girls from sexual exploitation, the bishop's inclination was to restrict all measures to the 'really . . . practicable', and he candidly admitted in speaking for the bill that he had disagreed with the Associate Institute here.

Since the church was, notoriously, the landlord in many slum areas where brothels were based (Shaw's *Mrs Warren's Profession* is in a long line of comment on such anomalies) Philpotts may have been very happy to see the proposed penalties against the owners of broth-els thrown out in committee. He also emphasized in the House that the 'seducer' was not to be made liable for conviction. His justification was that the quest for legal redress would tend to turn a fallen girl aside from the path to 'the sacred, the hallowed rites of repentance', but it may be that criminalizing the seducer was one of the measures which Philpotts had persuaded the Associate Institute to keep out of the bill. Among the steps he labelled not 'practicable' was any 'undue inquisition into the habits of individuals'. If the Associate Institute did originally wish the new legislation to criminal-ize the libertine male—as I think is probable—they would have been thoroughly disappointed. Even a man acting as 'Walter' often reports of himself in *My Secret Life*, namely, paying a brothel-keeper to pro-cure a particular girl, could not have been touched by the 1844 bill—though his instrument could.[20]

Events seemed to be moving in something of an opposite direc-tion, however, with the next Associate Institute bill, broadly similar to that of 1844 and sponsored in the Commons by R. Spooner: the Seduction and Prostitution Bill of 1847. While references to brothel-keepers were removed in committee the scope of the bill was also enlarged precisely to net a Walter, or 'any person who shall pay, or offer or promise to pay, any money, fee or other valuable considera-tion whatever for the purpose of so procuring, or as a reward for hav-ing procured . . . such seduction as aforesaid'. It passed the Commons, but the Lords were still too nervous of the dangers it posed for innocent urban landlords and threw it out.

In the next session of Parliament Samuel Wilberforce, Bishop of Oxford—a soapily different figure, ideologically, from his Evangelical father, but also not as warm a spokesman for this legislation as Philpotts had been—sponsored the Protection of Females bill. This was the most restricted measure yet, targeting only procurement for gain, with the extra proviso that there should also be an element of deception involved in the offence. Even this was too ambitious for the Lords, and the bill only passed both houses when reintroduced shortly afterwards as a one-paragraph measure. Its scope was confined to anyone who 'shall, for the lucre of gain, by false pre-tences, false representations, or other fraudulent means, solicit and procure any woman to have carnal connexion with any man'. Five

---

[20] *Hansard*, 74 (1844), cols. 1233–4, 75 (1844), cols. 878–80.

years of effort by the Associate Institute to stir the moral energies of
Parliament had yielded a pusillanimous statute.

There is some direct evidence about the sexual-moral convictions
of Wilberforce's principal associates in the first Evangelical campaign. It is apparent that, within this pioneer group, environmentalist
attitudes towards human sexuality had made headway, even if the
practical face of early Evangelicalism on these questions was gov-
erned by traditional notions of innate depravity. Most strikingly,
Thomas Gisborne accepts a *tabula rasa* account of the moral condition of a young girl, only entering a qualification which is really no
qualification at all: 'The human mind in infancy has been compared,
and in some respects justly compared, to a blank sheet of paper. In
one material point, however, the comparison fails . . . If you forbear
to impress it with ideas and impressions, can you prevent it from
receiving impressions from the persons and objects with which it is
daily conversant?'[21]

To Hannah More belongs the distinction of having written at
greater length explicitly about sex than any other leading
Evangelical. Her *Coelebs in Search of a Wife* (1809) is as tedious an
experience as a three-decker novel devoted entirely to a very virtuous young man's search for a very virtuous wife is bound to be. It
could have been worse, for Hannah More never writes badly. And it
is not completely unerotic. This would have been hard to achieve,
given the nature of the hero's quest, but William Godwin in *Caleb
Williams* at least shows how an author of truly anti-sensual views
could depict an energetic and morally unsupervised young man who
has not a thought in his head for the other sex (there are some
intriguing resemblances, in fact, between Hannah More's novel and
Godwin's story of Caleb in search of a secret, of fifteen years earlier).
Coelebs exhibits a respectable degree of distress at the idea of having
to wait three months before he can marry his selected bride, Lucilla
(so that she will be 19)—and his frustration is comprehensible in the
light of the many allusions to Milton's Eve in Paradise as the type of
the ideal wife. The novel ends on an equivocally anti-sensual note
somewhat reminiscent of Rousseau's *Emile*, in the middle of this
period of premarital deprivation.

Broadly speaking, *Coelebs in Search of a Wife* allows the stereotypical experiences of young love to remain intact in the new framework

[21] Gisborne 1797: 44–5.

of a quest by the ultra-virtuous for the ultra-virtuous. This is especially striking in view of remarks uttered by one of the characters in
the novel, Mrs Stanley, about the way fiction has fostered misconcep-
tions about sexuality:

Mrs Stanley lamented that novels . . . had done infinite mischief by so completely establishing the omnipotence of love, that the young reader was
almost systematically taught an unresisting submission to a feeling, because
the feeling was commonly represented to be irresistible . . . ill-placed affection only became invincible, because its supposed invincibility had first been
erected into a principle.[22]

Coelebs's affection for Lucilla is emphatically not 'ill-placed' but nor,
by a less stringent measure, is the passion described in the hundreds
of novels of the period which issue in happy marriages. So one might
say that *Coelebs in Search of a Wife* and many of its less Evangelically minded peers have in common the invincibility of well-
placed affection.

The interplay of the artefactual and the authentic in sexual drives,
with the moral implications to be drawn, was evidently a topic which
interested Hannah More, and she wrote about it quite subtly. In her
Strictures on the Modern System of Female Education (1799) it is
chiefly German plays which are the mischievous agency. Writings of
this sort,

effect the most desperate work of the passions, without so much as pretending to urge their violence in extenuation of the guilt of indulging them. They
solicit this very indulgence with a sort of cold-blooded speculation, and
invite the reader to the most unbounded gratifications, with all the saturnine
coolness of a geometrical calculation . . . This cool, calculating, intellectual
wickedness . . . descants on depravity as temperately, and details its grossest
acts as frigidly, as if its object were to *allay* the tumult of the passions, while
it is letting them loose on mankind.[23]

Two interesting implications are evident here: (1) that sexual drives
are very strong and may be aroused into vigorous activity by even the
most neutral mention of sexual topics; (2) the strength of these drives
is a mitigation of sexual transgression when it is truly passionate.

The latter is one of Hannah More's habitual if unexpected themes.
The unforgivable sin is to have a lax *code*, rather than a lax nature:

that our passions should be too often engaged on the side of error, we may
look for the cause . . . in the unresisted propensities of our constitution: but

[22] More 1830: vii. 164.   [23] Ibid. v. 33.

that our *reason* should ever be exerted in its favour . . . that our *judgement* should ever look on with indifference . . . this has no shadow of excuse—because this can pretend to no foundation in nature, no apology in temptation, no palliative in passion.

And not only is there a good chance of recovery from sexual depravity as long as it has been motivated by passion, but, it is perhaps hinted, the very experience of transgression may build a more resilient moral being:

though the evil propensities of fallen nature, and the bewitching allurements of pleasure, will too often seduce even those of the best education into devious paths, yet we shall find that men will seldom be *incurably* wrecked unless that internal corruption of principle has taken place, which teaches them how to justify iniquity by . . . the sanction of false reasoning.

Those who are driven by unresisted passion into irregularities which shock their cooler reason, are more liable to be roused to a sense of their danger than persons whose perceptions of evil are blunted through a round of systematical, unceasing and yet not scandalous dissipation.[24]

The code, the core of intensely felt Christian principle, is all. Thus Wilberforce can be a Jeremiah in the House of Commons about modern sexual morals, while conceding the great 'moral decency' of the times. By the same logic Hannah More finds herself raising suspicions about the new spirit of philanthropy abroad in the upper classes at the close of the eighteenth century.[25] But how could a movement which put such a great emphasis on the inner integrating core of belief, without which no decency or philanthropy counts—and this is indeed the cardinal thought of the Evangelical revival—have also attached such importance to external observances? It is this familiar ambiguity which tends to produce nonsense when a snap definition of Evangelicalism is called for: did it put 'faith before works', or 'works before faith'?

In its practical effects this puzzle did not matter too much—otherwise the Evangelical movement could not have achieved its enormous practical success. Inner faith is an imponderable, or certainly not such an unmistakable entity that one can in the last resort ban a piece of 'decent' behaviour in the confidence that faith is missing, or condone a piece of vice in the confidence that it is present. In practice, one has to take the outer tokens of the inner spiritual condition as reliable—so that the distinction between genuinely Christian

---

[24] Ibid. v. 325, 1833–4: ii. 265, 315.　　[25] More 1830: xi. 87–8.

behaviour and the shallow parroting of Christian behaviour tends to collapse. The upper and middle classes are inclined to have right outward behaviour in place anyway, so here the effort must be to raise the inner Christian consciousness. The lower classes definitely have the wrong outward behaviour in place, so it is appropriate, as well as more straightforward, to put a special emphasis on its improvement.

This led to one inconsistency in the practical handling of moral questions by Evangelicalism: its well-known leniency towards upper- and middle-class backsliding, and severity towards the working-class equivalent. But even on working-class moral culture the Evangelicals did not have a single policy: there was still room to put a greater or lesser emphasis on inner spiritual change. The Vice Society had virtually no concern with the inner correlates of external compliance. The extreme magdalenists in the prostitute rescue movement, by contrast, believed in the capacity of even the most morally degraded individuals to transform their beings through Christian faith, and attacked sin by this route as much as through the control of prostitutes' regime, demeanour, and dress.

There was a second issue which complicated Evangelical policy on the working class, namely, the role of philanthropy. The main thesis of Ford K. Brown's *Fathers of the Victorians* may be distilled as the claim that the philanthropic aspect of Evangelicalism was a fake. All humanitarian and philanthropic ventures—even the campaign to abolish slavery—were judged in the light of the great ulterior motive, the hidden agenda of Evangelicalism: the spread of Christian commitment. It is true that all Evangelicals were wedded, by definition, to the overriding importance of right belief: it was certainly not worth gaining the whole world to lose one's soul in this sense. Hannah More thought that philanthropy was valueless without Christian commitment both for donor and recipient, and her celebrated Cheap Repository stories routinely extol the combination of indigence and piety.

It is also true that in Hannah More's tales, including the most famous of them all, *The Shepherd of Salisbury Plain*, some improvement of the material circumstances of the indigent faithful does come about. The early Evangelicals gave to charity from their personal wealth on a very large scale. Within the orbit of Evangelical activity came several ventures—notably the Bettering Society, the Philanthropic Society, and the Juvenile Delinquency Committee— which, as I have explained in the first part of this study, were envi-

ronmentalist in their approach to working-class morality: finding
little if any place for notions of a vitiated moral core in the individual.
The leaders of first-generation Evangelicalism did not exert themselves greatly in these projects, but they had the blessing of
Wilberforce and others. It is arguable that the Evangelical belief in
conversion, or rather in the redeemability of the soul which is the
basis of conversion, came together here with the philanthropic
impulse, to yield an interesting compound philosophy: that souls may
be spiritually improved by a change in social conditions.[26]

And if some of their Evangelical sponsors regarded these projects
as being of secondary importance, not all workers could have had the
clear vision of ultimate religious goals of a Wilberforce or a Hannah
More. The breadth and depth of the Evangelical venture as a whole
made a homogeneity of purpose impossible. 'There is scarcely one
Englishman in ten', wrote Macaulay, 'who has not belonged to some
association for distributing books, or for prosecuting them; for send-
ing invalids to the hospital, or beggars to the treadmill; for giving
plate to the rich, or blankets to the poor'. The coexistence in
Evangelicalism of rather incompatible strains of philanthropy and
spiritual concern is brought out very nicely in Macaulay's paradoxes.
Ford K. Brown concedes that in the rank and file of Evangelical sub-
scribers, committee-members, and officers 'probably there were
many' who were 'men and women of simple goodheartedness and
unostentatious piety', 'more concerned with what they took to be
good Christian conduct and less with points of view of any kind'.[27]

Such people were drawn into active Evangelical work because
they perceived the movement as a great humanitarian agency. There
was above all the evidence of the campaign against slavery (if non-
cynically regarded), but also the work on behalf of chimney boys and
children in factories. With all this on one side of the balance, and distrust of Evangelicalism's religious and political affiliations (whether
the latter were seen as jacobin or conservative) on the other, it is not
surprising that there developed the 'Wilberforce enigma'. There was
both admiration and caution in many minds where the effective
founder of the Anglican Evangelical revival was concerned. Broadly
speaking, he was admired where he was progressive, and distrusted
where he was repressive.[28]

[26] F. K. Brown 1961: 89–90, 347–51, 510–12; Quinlan 1941: 133–4.
[27] F. K. Brown 1961: 317, 400–1.   [28] Howse 1953: 129–34.

II

Magdalenism

In June 1864 A. J. Munby recorded in his diary his reaction to a strange sermon preached by the maverick high church minister Joseph Lyne. Munby's main impression was, 'more than ever before, the close rapprochement of the extremes of Calvinism and Catholicity'.

This man was rudely humorous like Latimer: vehement like Knox; plain-spoken and forcible like the preaching friars: he asked his elegant fashionable hearers if they had not committed fornication last week; asked the ladies why they had not been helping prostitutes out of sin, instead of decking out their own contemptible bodies, as if they were not worse, instead of better, than the prostitutes.[29]

The episode helps to throw into relief the oddity of that distinctive element in the religious life of nineteenth-century England: the attempt to rescue prostitutes. How does it come about that an extreme Anglo-Catholic like Lyne (he set up a Benedictine order, unsanctioned, within the Church of England) can make it a serious reproach against fashionable women in his audience that they do not engage in such work? What can he mean by the suggestion that his hearers are worse than prostitutes? What verdict on the sexuality of prostitutes does this imply? How did women of superior rank receive such sentiments (Munby says that in this instance they were 'too wellbred to show any signs of emotion, whether of repentance or disgust')?

The resemblance between 'Calvinism' and 'Catholicity' which Munby notices might reduce the oddity of this moment somewhat. The standard picture of the development of a more sympathetic and personally engaged attitude towards the prostitute in the nineteenth century (which I shall call magdalenism) is that this was an Evangelical and Nonconformist initiative which in due course recommended itself to a wide range of religious opinion. Hence the *Contemporary Review*, a centrist Anglican periodical, was proud to speak up as follows, in 1870, for Christianity's treatment of the prostitute: 'it is religious philanthropy alone, with its deeper feeling of

---

[29] D. Hudson 1974: 197. For Lyne and his impact on women see also Davies 1873: 260–4, 1875: 163–70; S. Hardy 1992: 175–85.

duty, its more daring hopefulness, its more enthusiastic charity, which ventures upon the arduous path . . . following its Master into the wilderness, to seek and to save that which is lost'.[30] The reference is clearly to the activity of urban missionaries in seeking out prostitutes on the street and even in brothels, work which Evangelicals and dissenters in the 1850s had pioneered.

Their approach is not, in fact, completely endorsed by the *Contemporary*—which still, for example, sees some utility in prostitute registration under the Contagious Diseases Acts as offering an occasion for purely secular 'escape' from prostitution for some girls— but the main point is to draw a contrast with 'modern secularism' and its allegedly unfeeling and inhumane attitude towards prostitutes. The *Contemporary* quotes a strikingly sour remark by John Morley in the magazine he edited, the *Fortnightly*, about 'sentimental persistence in treating permanently brutalized natures as if they still retained infinite capabilities for virtue'.[31]

But here complications start to reappear, for 'secularist' opinion in the strongest sense tended to sympathize with the project of prostitute rescue. Indeed at exactly this date another magazine editor, John Chapman, in the course of three important articles in the *Westminster Review*, was finding—and deploring—a very similar pessimism and indifference about prostitution in the mass of Christian moral opinion and 'Christian purism'. It is true that Owenites, for example, disliked Evangelical rescue work and deplored prostitution, but this was bound up with their more comprehensive attack on capitalism, which they saw to be the cause of the whole problem in the first place. The great point with progressive opinion, and why it is potentially in tune with Christian reformism, is not to be defeatist about prostitution, or to make arrangements to accommodate it as an inevitability. The endpoint of the latter process is prostitute licensing, which secularists generally opposed. Richard Carlile's followers did argue for French-style prostitute regulation, but as a makeshift in a bad social system. Many years later John Stuart Mill spoke strongly against the Contagious Diseases Acts to the relevant parliamentary commission.[32]

---

[30] 'Regulation—Cure—Prevention', *Contemporary Review*, 14 (1870), 220–35.

[31] *Fortnightly*, NS 7 (1870), 374.

[32] *Westminster Review*, NS 36 (1869), 179–234, NS 37 (1870), 119–79, 477–535; Walkowitz 1980: 35; *Newgate Magazine*, 1 (1825), 441, 2 (1826), 410; *New Moral World*, 9 (1841), 97, 10 (1842), 151, W. H. Smith 1839: 47.

Morley's sentiments are something of a throw-back to an earlier time in the century, when one can easily encounter the attitude which declared that prostitutes were moral lepers, and the attempt to rescue them either ludicrous or corrupt. This point of view came naturally to the more racy guides to London life aimed at the male consumers of prostitutes' services—who probably did not mind being told that they were being preyed upon by a quasi-criminal sisterhood, rather than anything resembling the reverse. The founding in 1812 of one of the first rescue organizations in the century, the Guardian Society, was made the occasion for an attack on the whole project (the current scale of which is much exaggerated by this anonymous author) of trying to rescue prostitutes: the women on the streets, above all, are 'mortified members of the body-moral of this country, which ought to be cut off'.[33]

When a Methodist minister from Southampton, James Crabb, tried to set up a penitentiary in 1823 he encountered opposition both from Anglicans hostile to dissent and from non-sectarian philanthropists who were sceptical of the value or feasibility of prostitute-rescue. In the 1830s there was allegedly a story being guffawed at appreciatively by middle-class men about a girl who goes to an asylum saying 'I am out of work, cold, hungry, tired, houseless, and anxious to be saved from evil courses', only to be turned away as 'not being qualified'.[34] But it may be significant that this was being laughed at 'in private' by this date. It became less acceptable, as Victoria's reign wore on, to appear callous about prostitutes. Perhaps the last substantial discussion of the subject which is unsympathetic to the prostitute is Ralph Wardlaw's *Lectures on Female Prostitution* of 1842, though such sentiments will be found cropping up more transiently for several decades thereafter, and the examples given towards the end of the last chapter could be supplemented.[35]

The story of the not 'qualified' prostitute is in fact only indirectly callous about the girl, and its main target is the supposed nature of the prostitute penitentiaries of the period. As we shall see, rescue workers at the time shared this disquiet about the operation of the existing institutions. But when they started a programme of more personal and compassionate action around the mid-century anti-Evangelical opinion felt no less obliged to denounce the new kind of

[33] Badcock 1828: 64–72; *London Guide* 1818: 29–34, 114–16; 'S.T.' 1818: 233.

[34] Rudall 1854: 100–2; Wakefield 1833: i. 77.

[35] e.g. *British Controversialist*, NS 1 (1859), 256.

rescue work. There was, it was said, 'a kind of joint-stock depository' of hostile opinions on what was being attempted, 'much resorted to, by a large section of the community'. From now on this is the context of much of the unsympathetic comment on prostitutes, such as that in the *Saturday Review* or John Morley's remarks in 1870.

There is a second important shift in the nature of the debate at this time. Medical opinion, as expressed in the *British and Foreign Medico-Chirurgical Review* and the *Lancet* in the late 1850s, seems at once disgusted and moved by the thought of the prostitute, and ambivalent about attempts to bring her back into the fold, in a fashion which surely derives from the generality of professional middle-class opinion.[36] But both these organs were here also responding to a new view of the prostitute which did claim to have specifically medical authority, for it was set out in William Acton's *Prostitution* (1857). From this point on nineteenth-century arguments against prostitute rescue, and in favour of a degree of indifference to prostitutes, are perhaps all to some extent couched in Actonian terms, and given a new self-confidence by Acton.[37]

His account of the prostitute's life, which was itself indebted to the great French pioneer Parent-Duchatelet, struck at magdalenism in two major respects: the prostitute was not in a state of miserable remorse, and she did not need saving, because she would save herself in due course and marry respectably (and, for good measure, her life while a prostitute did not damage her health). Acton's prostitute was a happy and healthy hooker, literally 'gay' (to use the main slang epithet for her way of life), and after Acton this is the image which becomes the negative pole of ambivalence in the commentators. A neutral surveyor of prostitution, Bracebridge Hemyng, assimilates it into his classifications in the 1861 edition of *London Labour and the London Poor*:

The happy prostitute, and there is such a thing, is either the thoroughly hardened, clever infidel . . . who in the end seldom fails to marry well; or the quiet woman who is kept by the man she loves . . . The sensitive, sentimental, weak-minded, impulsive, affectionate girl, will go from bad to worse.[38]

---

[36] 'Phases of London Life', *British Quarterly Review*, 35 (1862), 341–66; Hale 1809: *passim*, 1812: *passim*; J. Evans 1809: *passim*; *Moral Reformer*, 3 (1833), 99, 307–9; *Magdalen's Friend*, 2 (1862), 130; *Lancet*, 1857(2), 478–9; *British and Foreign Medico-Chirurgical Review*, 21 (1858), 388–415.

[37] Nead 1988: 148–9.   [38] Henry Mayhew 1861–2: iv. 236.

But the 'impulsive, affectionate girl' on a downward trajectory from a sexual fall which has occurred out of love—who in 1861 stands in the gallery of prostitutes with the calculating and robust Actonian type—needs also to be noticed. She is a traditional figure not only with magdalenists but also to some extent with their opponents, and this adds another complication to the story of nineteenth-century sympathy for prostitutes. The school of thought which regarded prostitutes as 'mortified members of the body-moral of this country' did not, at least when it was serious, wish to condone the male sexuality which was responsible for prostitution. William Paley (although, as we shall see, he sought to undermine the theological foundations of magdalenism) was very harsh on what he called 'seduction', and this rather capacious notion has a remarkable currency in early nineteenth-century discussions of sexual morality.

The *Scourge* in 1814 carried a skit parodying various points of view on the seduction of a young girl. 'Moralist' seems to take a Benthamite line, 'Philosopher' a Malthusian one, and 'Sentimentalist' sees the victim as completely innocent, the seducer as completely black. This exaggerates the range of voices which felt some need to utter a condemnation of 'seduction', but not unduly. Certainly most shades of religious opinion had something to contribute on this supposed menace, including even the liberal Unitarianism of W. J. Fox. The *Scourge*'s author 'Proteus' professes no view, and classic moralism was indeed an enemy of all the attitudes caricatured here, including that of 'Sentimentalist'. Many years later the *Saturday Review* is still denouncing, as the fantasy of 'lady doctors and Dissenting ministers', the notion that every girl who falls is 'the victim of the most insidious and pertinacious corruption, to which she yielded only after the most stubborn resistance': 'the fact is, that for one case of one-sided seduction there are fifty cases of mutual and consistent unchastity'.[39]

Nevertheless, 'Sentimentalist's' conviction of masculine depravity and/or feminine innocence in a woman's sexual fall is surprisingly widespread with the same conservative opinion which condemned prostitute rescue work. This may not seem paradoxical, for 'seduction' (of Jane Austen's Lydia Bennett by Captain Willoughby, let us say) sounds very separate from prostitution. But 'seduction' was a

---

[39] 'Verbal Morality', *Scourge*, 8 (1814), 109–15; *Monthly Repository*, 7 (1833), 347–53; 'Seduction and Infanticide', *Saturday Review*, 22 (1866), 481–2; cf. *Saturday Review*, 20 (1865), 172.

hydra-headed notion whose full range of use in early nineteenth-century controversies requires a larger survey than is possible here. Partly it was indeed a way of giving expression to fears about adultery or other transgressions within middle- and upper-class life which were never likely to lead to prostitution. But in other applications it was closely connected to the latter. As we have already seen, the problems of devising legislation against prostitution which had any chance of appealing to Parliament led to an emphasis on 'seduction' in a brothel or similar setting, in other words the persuading of a girl to turn prostitute, as the central offence.

Most important, seduction in the more ordinary sense of the term was commonly believed to be a major precursor of prostitution. No doubt the commentators who took this line differed somewhat as to the kind of transaction they had in mind—as to the degree of social inequality between the parties, in particular—but they agreed that it had the effect of injecting into the prostitute population an especially pitiable, unhappy, and unjustly stigmatized element: in other words, magdalens. Whether this was a well-grounded belief or a device to rationalize ambivalent feelings about the prostitute one cannot know, but it certainly had the effect of allowing two different attitudes to coexist in several early nineteenth-century discussions, at once punitive and reformist, contemptuous and compassionate.[40]

One gets the sense at this date, in fact, of quite a vigorous but relatively secular vein of magdalenism, which commonly grounded its response to the prostitute in the fact of 'seduction'. It had something of the same worldly-wise constituency as classic moralism (one of its representatives is the Regency buck George Hanger), but its empiricism pushed it in an opposite direction.[41] And a line of descent from this position can be traced in the Victorian years, sometimes in the form of proposals for a less severe regime of rehabilitation than is offered by the ordinary penitentiary or asylum. Dickens, planning Urania Cottage with Lady Burdett-Coutts in the 1840s, was not the only man thinking along these lines. In the 1860s secular magdalenism even gathered steam. Mainstream women's fiction becomes discernibly more sympathetic towards fallen women than it had been in the previous two decades. The *Athenaeum* perceived the more under-the-counter *Anonyma* series of novels, with their remarkably unflustered depiction of courtesanship and the fast life, as an

---

[40] *Hints* 1811; Kidd 1837: 32–5; 'S.T.' 1818; F. Smith 1809: *passim*.
[41] Hanger 1801: ii. 250–76; 'Amicus Patriae' 1810: 51–3.

aftermath of society's taking up the problem of ordinary prostitution and making it 'the fashionable topic of the London season'.[42]

One way of judging the role of this more-or-less secular magdalenism may be expressed by saying that the fashionable auditors at Joseph Lyne's sermon in 1864 would not have felt nudged in the direction of a greater kindliness towards prostitutes only by the exhortations of this eccentric Anglo-Catholic preacher. The possibility might have already been on their minds as women-of-the-world living in London, reading novels, and so forth. If their allegiance to the Church of England was very tenuous they might even have experienced magdalenist urges of the utilitarian and progressive sort which John Chapman gives voice to.

The important point survives that the exhortation to *practical* magdalenism comes from a religious source. Secular kindliness to prostitutes in the period was not much more than a velleity, perhaps taking concrete form occasionally when a magistrate or police officer helped a prostitute registered under the Contagious Diseases Acts to give up the life, or in a donation to one of the rescue institutions (and the latter were chronically short of funds). Patrick Colquhoun suggested—with support from *The Times*—that there should be a network of publicly financed penitentiaries in London, and there are other scattered proposals for apparently secular operations of this sort,[43] but Dickens's Urania Cottage, funded by Angela Burdett-Coutts, seems to be the only such institution which was not run by religiously committed individuals—and even at Urania religious teaching had a central role. John Chapman's magdalenism takes the form of a reproach to modern orthodox Christianity for not being true to its principles. In the nineteenth century the proposal that prostitutes should be treated in a sympathetic and supportive way is mainly a religious initiative even at the level of theory, and almost entirely so at the level of practice. To this extent the whole phenomenon partakes of the oddity which comes out in Munby's anecdote.

When the *Contemporary Review* speaks of Christianity 'following its Master into the wilderness' on prostitution, the primary reference is probably to Christ's encounter with the woman taken in adultery,

---

[42] Wade 1859: 388–90; S. Mitchell 1977; *Athenaeum* (22 Oct. 1864), 523. See also Skene 1865: 5.

[43] Bristow 1977: p. 66. See *Thoughts on Means of Alleviating the Miseries attendant upon Common Prostitution* (London, 1799) and Mary Ann Radcliffe, *The Female Advocate* (London, 1799).

at John 8: 3–11. She is brought before Jesus by scribes and pharisees, who challenge him to endorse the Mosaic punishment of stoning for her offence. By putting his endorsement in the form, 'he that is without sin among you, let him first cast a stone at her', Jesus discomforts the woman's accusers, and they depart.

And Jesus was left alone, and the woman standing in the midst.
. . . he said unto her, Woman, where are those thine accusers? Hath no man condemned thee?
She said, No man Lord. And Jesus said unto her, Neither do I condemn thee: go, and sin no more.

It is one of the most memorable and satisfying anecdotes in the Gospels, though ambiguous as to Christ's teaching on adultery and attitude towards the woman. The sense of a Jesus who was positively lenient towards female sexual transgression comes more from the traditional interpretation of the 'woman of the city, which was a sinner', with 'sins, which are many' (her story is told at Luke 7: 37–50) as a prostitute. She washes Christ's feet with her tears, wipes them with her hair, and anoints them with ointment from a box of alabaster. It is not actually said that she is penitent, but Jesus tells her that her sins are forgiven, 'for she loved much'.

In Christian tradition the motifs of this woman's loving nature, and Christ's forgiveness of her sins, are strengthened by her identification with two figures called 'Mary' in the Gospels: her name supposedly appearing variously as 'Mary of Bethany' and as 'Mary of Magdala' or 'Magdalene' (out of the latter Christ drives 'seven devils' at Luke 8: 2). The connection with the 'woman of the city' is via John 11: 2, where Mary of Bethany is called 'that Mary which anointed the Lord with ointment, and wiped his feet with her hair'. It appears to be Mary of Bethany who 'sat at Jesus' feet, and heard his word', instead of helping her sister Martha, at Luke 10: 11. Mary Magdalene, of course, is a privileged witness of the resurrection in three of the four Gospels.

The tenuous chain of biblical association between this Mary and sexual sin has, traditionally, been strong enough for her name to denote the motif of the penitent and forgiven prostitute. But it does not manage to include the one certain encounter between Jesus and a female sexual sinner, namely, the woman taken in adultery. Moreover *this* figure is not said to be penitent, or affectionate, and Christ does not precisely forgive her, let alone take her into his following. William Paley, in a statement of Anglican doctrine on sexual morality which cast its shadow well down the nineteenth century, denied that adultery was forgiven by Christ in this episode. The Gospel only says that Christ did not 'condemn' the woman, which in context, as Paley points out, might mean no more than 'find guilty'.[44] The technically innocent adulteress thus conjured up by Anglican theology is a far cry indeed from the magdalen of tradition: that passionate and abjectly remorseful woman whose capacities for love are transfigured as rapt discipleship.

The Evangelical and Nonconformist camp which promoted prostitute rescue was aware of the fallacy of calling prostitutes 'magdalens', but indifferent to the problem—indeed breezily indifferent given the Roman Catholic associations of this usage and its lack of biblical sanction. 'We cannot be held accountable for a popular error . . . nor refuse to make use of a word in its accredited and popular sense'. A licence with scripture was apparently, in this instance, 'accredited' by a certain image of the prostitute. It would however be a simplification to suggest that 'Evangelical' opinion in a broad sense— zealous low church Anglicanism and Nonconformity—was solidly in favour of magdalenist policies. There is evidence that Quakerism, which has not figured hitherto in my survey of religion and sexuality, was an important trigger for the more active phase of missionary work with prostitutes, both practically and through the model of Elizabeth Fry's efforts for female prisoners. The Quaker Jonathan Dymond argued that prostitution would be much reduced if a respectable woman's fall were treated as 'venial'. At the York Refuge in the latter half of the century Quakers appear to have been squeezed out of the administration of the home. Conversely the up-market dissenting magazine, the *British Quarterly Review*, was in 1862 still grudging in its admission that there was a core of 'indestructible humanity' in the prostitute. Dissenting opinion had in fact always been divided between the options of compassion and severity towards prostitutes. The work of the London Female Penitentiary and similar bodies in the first years of the century was both attacked and defended publicly by writers known to be, or apparently, dissenters. The Baptist founder of teetotalism, Joseph Livesey, was a notable advocate of kindly treatment for prostitutes and of female involvement in rescue work.[45]

[44] W. Paley 1785: 257–9.

[45] *Magdalen's Friend*, 1 (1860–1), 63; Bristow 1977: 67; Dymond 1829: i. 379; Finnegan 1979: 171; Hale 1809: *passim*, 1812: *passim*; 'Juvenis' 1809: *passim*; J. Evans 1809: *passim*; *Moral Reformer*, 3 (1833), 99. See also Kruppa 1992.

But whatever the divergences within the Evangelical and dissenting world on the issue, it was individuals and groups of individuals from this world who launched prostitute rescue work in nineteenth-century England, who dominated it for half a century, and who made the running for as long as this kind of work went on. A solitary Anglican enterprise, the Magdalen Hospital in St George's Fields, founded in 1758, was for almost forty years the only institution in England seeking to rehabilitate prostitutes, until Thomas Scott, an early Anglican Evangelical and chaplain at the Lock Hospital, created the Lock Asylum in 1787. While the Magdalen remained the only Anglican agency of any consequence in the field until the 1840s, with an apparently declining intake of penitents,[46] Evangelical and dissenting initiative was setting up a dozen or so metropolitan and provincial operations, including the London Female Penitentiary (1807), the Guardian Society (1812), the Metropolitan Female Asylum (1829), the British Female Penitent Refuge (1829), the London Female Mission (1836, later the Female Aid Society), and in the provinces penitentiaries in Bristol (1802), Liverpool (1811, supplemented by the Benevolent Society for Reclaiming Unfortunate Females in 1838), Leeds (1821), and Manchester (1822). As far as the evidence goes, these institutions were all under firmly religious direction. The one identifiable exception is the Metropolitan Female Asylum, founded in 1829, which made some effort to keep sectarian religion out of its administration.

The Church of England was finally jolted into new prostitute rescue work by the success of the Catholic Good Shepherd asylum in Hammersmith, an example to which the growing high church party of the 1840s was particularly responsive. The purely private efforts of individuals such as the young Gladstone, the founding by Pusey of Anglican sisterhoods, and the concrete proposals of John Armstrong, came together in 1849 with the creation of the first sisterhood-based Anglican penitentiary at Clewer, near Windsor, with Gladstone on the committee. Within ten years there were some twenty such asylums associated with lay sisterhoods (in other words, with almost all that had been formed), linked through the Church Penitentiary Association, admitting something over a hundred women a year. Though the model was distinctly Catholic (and, more specifically, that of the French *Sœurs de la Charité*), the Anglican penitentiaries

---

[46] Manning 1844: 35; T. T. Carter 1861: 15.

were presented by their sponsors to the broad public as an important improvement on the now familiar Evangelical penitentiaries in point of humaneness and concern for the spiritual improvement of their inmates.[47]

At very much the same date Charles Dickens was conceiving a small non-denominational asylum in London which, likewise, would contrast favourably with the usual type in its regime and attitude to its inmates. Urania Cottage opened in 1847, and remained in operation for something over ten years.[48] But in the mean time, rescue workers of an Evangelical persuasion were stealing a march on these would-be reformers with innovations of their own. In the latter 1840s a retired naval lieutenant of burning religious convictions, John Blackmore, became dissatisfied with the strict and prison-like character of the penitentiary with which he was associated (probably one of the London Female Mission's homes), and founded his own pair of homes, one of which, the London Female Dormitory, was open to all applicants. Inmates were actively recruited on night-time, or 'moonlight', expeditions by Blackmore and a colleague.[49] Blackmore's homes set the precedent for a new type of 'family' establishment, structured on domestic lines. The Rescue Society, under Daniel Cooper, from the early 1850s admitted prostitutes' children along with their mothers.

In 1849 Blackmore, with Theophilus Smith, had experimented with another technique for drawing possible penitents, in the form of late-night meetings, essentially religious revival gatherings, to which prostitutes were invited by cards distributed on the pavements. The idea was abandoned but revived with considerable success in the West End and other London localities ten years later. The Midnight Meeting Movement spread to other major cities, and seems to have been prosperous for about eight years, though it slackened thereafter.[50] In London it used in particular the services of three organizations as receiving agencies for penitent prostitutes: the Rescue Society (founded in 1853), the London Female Preventive and Reformatory Institution (founded in 1857, and one of the most durable and extensive of these enterprises, with four homes), and the Homes of Hope, founded in 1860.

[47] Goodman 1862: 253; R. Knox 1857: 188; T. T. Carter 1853: 8, 1861: 7, 19; J. Armstrong 1848, 1858: 50; *Morning Chronicle* (8 August 1851), 4, (18 Sept. 1851), 5, (22 Sept. 1851), 5.
[48] Collins 1964: 111.
[49] Blackmore 1860: 19–24, 34–8, 1889: p. vi; Heasman 1962: 150–1.
[50] *Statement* 1860: 4, 16–37; *Friend of the Fallen*, 1 (1870), 12; Heasman 1962: 152.

Just as the Midnight Meeting Movement was being revived another novelty in rescue work was launched, namely female missionary work in the sites of prostitution. In 1858 the Female Mission to the Fallen was set up, in London, by the Reformatory and Refuge Union, an important umbrella organization of which Shaftesbury was president; it became financially self-sufficient in 1871. It was a natural outgrowth of female evangelizing among the urban working class (and missionaries had for some years made occasional use of the penitentiaries); it was in tune with, or directly influenced, initiatives by women in other centres, such as Julia Wightman in Shrewsbury, Josephine Butler in Liverpool, Sarah Robinson in Aldershot, and Ellice Hopkins in Brighton. In Ellice Hopkins's case, the sense of the flagging of the Midnight Meeting Movement, and of the need for a new approach, was strong.[51]

The exact number of rescue organizations, at any time, has not been determined; of some there may now be no surviving record. The Evangelically minded institutions were more vociferous about their activities, individually and as a group, than were those with other persuasions, and there is a reliable figure of sixty such operations, throughout the country, in 1860. A few of these ran more than one home. By this date there were also about twenty Anglican penitentiaries, and certainly some Catholic ones. But the most impressive growth in rescue work in a sense occurs after this date, for the pace of the last seventy years was kept up, if not accelerated, and virtually all religious groups, including the Jewish, seem eventually to have been caught up by it. There were at least 308 asylums in being in 1906, even though prostitute numbers had not risen remotely in proportion.[52]

I shall mention numbers of penitents in more detail later, but it is important to indicate the scale of the asylum accommodation proper (bearing in mind that there was also a good deal of transient shelter and succour being offered). Around 1860, I would judge, there cannot have been more than 2,500 places available in English asylums of all kinds. This is tiny compared with the higher estimates of prostitute numbers that were current, but much more substantial compared with the lower, and more convincing, estimates. There may well have been one asylum place for every six prostitutes in London. Thus the metropolis fell somewhere between Liverpool, which had

[51] London Female Penitentiary 1836: 13; Prochaska 1980: 189; E. Hopkins 1877: 2.

[52] *Magdalen's Friend*, 1 (1860–1), 94–5, 126–7, 190–1; Prochaska 1980: 188–90.

very scanty provision, and York, where there seems to have been one place for every three.[53]

A constant theme in the development of rescue work, at least on the lips of the rescuers, is improvement, in an ever more magdalenist direction. The bad old first generation of committee-run, prison-like refuges is allegedly replaced by more spiritual, or more accessible and welcoming, or more cheerful and informal, or more homelike establishments: with a correspondingly more compassionate and friendly attitude to the women who could or did enter them. New intentions are certainly signalled in the way the vocabulary of 'home', 'cottage', 'dormitory', 'mercy', and 'mission' replaces 'penitentiary' and related words. There is a great superficial difference between the group of intrepid respectable women who started to walk into London brothels in the 1850s, calling the girls 'sisters' and telling them that Christ could wash away the past, and arrangements at the ancestor of all nineteenth-century penitentiaries, the Lock Asylum, which by this date seemed repellently severe.[54]

But the historical outsider, at over a hundred years' remove, may well be struck by the persistence of certain themes in the running of these rescue operations, even from the time of the Magdalen. The founders of the London Female Penitentiary, which in 1840 was judged a leader among such places for 'benevolence, wisdom and lib- erality', had quite consciously looked to the Magdalen as a model when devising the asylum's rules in 1807.[55] When around the mid- century the third wave of rescuers set out to improve on established systems, they nevertheless retained a good deal of what they inher- ited. The problem for the historian is to judge the significance of the persisting elements. They may indicate a persistence of fundamental aims and attitudes concerning prostitutes, despite the rhetoric of greater kindliness and sympathy. Or they may be consistent with a real change of approach, being devices or procedures which could bear genuinely different meanings. They may also be playing a reduced role, even when they survive, so that the way the prostitute is regarded can be better judged by looking at the new kinds of res- cue activity with which they have been supplemented.

If the rhetoric of kindliness and sympathy is hollow, how is it to be explained? Was it just window-dressing, designed to attract victims,

---

[53] *Magdalen's Friend*, 22 (1862), 92–3; Butler 1928: 45; Bevan 1843: 19–20; Finnegan 1979: 167.

[54] Acton 1870: 193.   [55] Tait 1840: 253; London Female Penitentiary 1808: 8.

and financial sponsors, for processes which actually punished and humiliated prostitutes? Or in proposing this explanation do we fail to enter into a mode of thinking in which punitive and kindly intentions were felt, without hypocrisy, to be compatible? Or were the rescuers of prostitutes just muddled and ambivalent in their aims—and perhaps not guilty of any more fudging of purposes when they tried to state them than is often involved in our modern term 'rehabilitation'? The field of nineteenth-century magdalenist activity is virtually unexplored,[56] and much primary evidence still needs to be brought to light. I only offer the outline of an approach to these difficult questions, under the rubric of three of the persisting themes in nineteenth-century prostitute work: rescue, confinement, and selection.

Rescue: Much of the *frisson* of A. J. Munby's anecdote comes from the notion of a personal engagement with prostitutes on the part of respectable individuals, of the sexually non-transgressive with the grossly transgressive. This element becomes more pronounced in the magdalenist programme but it is there from the start, simply in the sense that magdalenists went predominantly down the route of reform of individual women—vetted, managed, and prepared for respectable life in the penitentiaries—rather than down the route of global suppression of prostitution. May one conclude that it was not simply 'the constant aim . . . to get the offending women out of the theatres and off the streets', as a hostile account has put it?[57]

Not necessarily. Global suppression of prostitution meant recourse to the law, which as it stood was not easily made effective against prostitution, and not easily reformed. Some prostitute rescue organizations, especially in the early days, did keep legal action and legal reform simmering as a supplement to the penitentiary—and sometimes as the preferred option. The Guardian Society, which had Wilberforce as vice-president, was 'guarding' public morals in the City of London, not young women. Its main goal, enshrined in a petition to the City authorities, was more prosecutions, its secondary goal 'to afford refuge and relief to such females, as . . . appeared only to need a temporary asylum to promote their reformation'. The notion of 'temporary asylum' looks forward oddly to the innovations of Blackmore forty years later, but one may suspect that its appeal in this context was cheapness, rather than sparing the prostitute the

---

[56] See however L'Esperance 1979.   [57] F. K. Brown 1961: 437.

rigours of full penitential discipline. Some members of the Guardian Society did regard the asylum in Bethnal Green (which admitted some two and a half thousand girls between 1812 and 1861, at a declining rate) as a device to 'remove' prostitutes from the streets.[58]

Another body which gave a primary emphasis to the law was the London Society for the Protection of Young Females, which soon turned itself into a plain 'Society' for these aims, and created an 'Associate Institute' to co-ordinate nation-wide demands for change in the law, while it brought many actions against brothels in the courts. It also placed girls in asylums, and opened its own when it found that girls under 15—the society's particular concern—were badly catered for in the existing institutions. This home failed to operate at full capacity for lack of funds, and took in between fifty and seventy girls a year.[59] Unusually, something of the general temper of the society can be deduced because its secretary, James Talbot, wrote one of the period's general treatments of prostitution, *The Miseries of Prostitution* (1844). This comes across as rather a bureaucratic compilation of alarmist data, with no strong magdalenist note.

The Protection Society was above all concerned with the law on brothels and procuring, and here it is understandable that rescue agencies regarded the law as a necessary adjunct to their work: 'in vain are all our Reformatories and Refuges . . . if the floodgates of vice be allowed . . . to remain open'. But as far as I know the police and the courts were not regarded as antagonists by prostitute rescuers in any phase of their handling of prostitution. The London Female Mission welcomed the enlarged powers proposed for the metropolitan police in 1839: prison was as good a lodging for the prostitute as the penitentiary. Under the Contagious Diseases Acts, somewhat paradoxically, the police acquired powers in relation to individual women which arguably had a benign potential: there were in fact some prostitute rescue workers who opposed the repeal of the Acts because they provided a mechanism whereby prostitutes could be referred to asylums.[60] But before this, and in general, there was on the face of it a possible clash between the police's function and

---

[58] Bristow 1977: 53, 67; Guardian Society 1816: 9, 32; *Reformatory and Refuge Union Journal*, 1–4 (1861), 70; Talbot 1844: 69.

[59] London Society 1839: 15; *The Times* (5 May 1846), 5; J. Armstrong 1848; Bristow 1977: 56.

[60] *Female's Advocate*, 2 (1839), 228–9, 3 (1840), 208–10; Prochaska 1980: 206–7. See also Rescue Society 1871: 10, Reformatory and Refuge Union 1872: 31.

that of the workers. One can see why the *Saturday Review* thought sectarian support for greater police efforts against prostitution betrayed the 'Puritanical prudery' of these groups.

It is disconcerting to find that some individuals who put considerable efforts into the full-blooded rescue work of later years were also active in purely suppressive operations. A case is Theophilus Smith, who pioneered the Midnight Meetings with Blackmore in 1849 but in 1866 tried to set up the West London Association for the Suppression of Vice to lobby for greater police powers of arrest where prostitutes were concerned. The Earl of Shaftesbury, patron of Blackmore's London Female Dormitory and instigator of the Female Mission to the Fallen, seems also to have founded in the 1860s a National Association for the Suppression of Public Immorality, through which the major receiving agency from the Midnight Meetings, the London Female Preventive and Reformatory Institution, added, or sought to add, a legal dimension to its work.

On a thoroughly cynical account, in fact, the Midnight Meeting—which in a matter of hours took prostitutes off the West End pavements, via tea and a sermon in the St James Restaurant, to a permanent or temporary asylum—was an inspired alternative to the police round-up, substituting the religious susceptibilities of young women for the bobby's arm. Its sponsors did boast that the pavements were visibly less congested by prostitutes while the meetings were taking place. Such an account would almost certainly be unfair to the motives of the workers who went on for many years organizing Midnight Meetings in the East End. Nor can it apply to the other major new technique for personal encounter with prostitutes developed at this period, namely brothel-visiting, as pioneered by Blackmore and John Vanderkiste and then taken up by female missionaries. Here the lay impression that 'repression having been abandoned for the nonce, the existing desire on the part of philanthropists seems to be to experimentalise on the laws of kindness' is more clearly justified.[61] John Blackmore seems to have been a man with a quite genuine ardour for the moral purification of prostitutes, rather than the salubriousness of the West End.

But even this kind of work could slide into suppression. An important co-worker with Blackmore, Edward Thomas, became secretary of the London Female Preventive and Reformatory Institution and

---

[61] Davies 1874: 364–71; Judge 1874: 31.

helped to lobby the police authorities on closing-hours in the West End. By 1879, at any rate, his magdalenism seems to be drained of all humaneness, and is exactly on a footing with legal suppression: 'amongst the most merciful measures which can possibly be adopted is . . . to exercise the sterner measures of the law so that they may be employed simultaneously with the milder means set forth in the Gospel of Christ'. This is indeed to strain the quality of mercy. The whole magdalenist movement was in a sense implicated at the end of the century in a suppressive approach to public morality, as its off-spring, the National Vigilance Association (founded in the offices of the Reformatory and Refuge Union in 1885), started to move the levers of the law in the fashion of the old Vice Society.[62]

This brings us to a core issue: how were prostitutes and their sexual promiscuity regarded by missionaries, and in particular by women workers? A fundamental point to make is that the Evangelical account of the prostitute does not, contrary to what one might expect, include a special theory about prostitutes' sexuality. It was common ground between rescue workers and most other commentators that women fell into prostitution for a variety of personal, social, and economic motives. It was also common ground that the last kind of motive, the economic, was very important, perhaps overwhelmingly so. But most writers conceded that an inclination for sex had an influence in some cases and, as we have seen, it was believed that a number of prostitutional careers started with an illicit surrender to passionate love (this becoming a disaster because of the treachery of the male and/or society's unforgiving code).

One could seek to show that the relative importance attached to these different motives depended on the ideology of the commentator, but this would be a laborious exercise and probably a futile one. Certainly it was no part of the Evangelical approach to deny that prostitutes were as they were, in some or many cases, because of their sexual drives. On the contrary: in so far as rescue work tended, as I shall suggest, to target women with respectable backgrounds— the group whose entry into prostitution was most likely to be attributed to a loving fall—the sexuality of rescue subjects was rather to the fore. Hannah More's remarks, quoted in the previous section, show that in classic Evangelical thought a woman's sexual susceptibility was no bar to her reformability.

[62] *Magdalen's Friend*, 2 (1862), 356; W. J. Taylor 1907: 81; E. W. Thomas 1879: 107, 165.

Charles Dickens is reported to have believed that 'God looked leniently on all vice that proceeded from human tenderness and natural passion', and in the publicity for Urania Cottage given to potential inmates he was at pains to avoid any sense that prostitutes were pariahs or moral inferiors: 'I address it to a woman . . . I write as if you were my sister'. Angela Burdett-Coutts, he says, has felt for prostitutes 'as being of her own sex and nature'.[63] It is characteristic of the whole magdalenist episode that Dickens's practice here is not as different from that of many sectarian workers as he probably believed it to be, and the problem is to know whether the two parties were doing the same thing for different reasons, or because of a basic agreement of aims. (I shall touch shortly on Dickens's tactic of making life at Urania somewhat more alluring than the reality, of 'tempting to virtue'.)

The rhetoric of non-condemnation, of a shared human nature, and of fraternity and sisterhood, is always on the lips of sectarian rescue workers from the 1840s—indeed literally on their lips, in face-to-face encounters with prostitutes. Baptist Noel, who became a leading speaker at the Midnight Meetings, would open his address 'dear young friends', and a female worker is recorded as making an impromptu speech to her 'sisters' at one of these gatherings. Women workers like Emma Shepherd and Ellice Hopkins urged their colleagues to behave as 'a woman among women' when in the field, and to be mindful of the prostitute outside on the street as they lay respectably in the matrimonial bedroom: 'yes . . . your *Sister*'. In a tendency which alarmed some workers, the prostitute even acquires a kind of superiority, as a woman coerced into pandering to men's callous lust simply because she once yielded to her own sincere passion. She is a 'beautiful and lovely woman . . . degraded, and perishing', not just a sister, but an enslaved sister, to be rescued by 'a kind of underground railway', and actually the moral better of the man who debauched her.[64]

The Christian truism of our 'all being sinners alike' has no doubt been put to some very inegalitarian uses, but one Midnight Meeting speaker, Reverend Goodhart, told his audience of prostitutes that it meant that he had 'the same wants, the same desires, the same

---

[63] Dickens 1981: pp. xii, 698.

[64] *Statement* 1860: 8; *Magdalen's Friend*, 1 (1860–1), 13, 121; E. Hopkins 1879: 35; Saint Mary-le-bone 1841: 12; La Touche 1862: 9; A. Macleod 1870: 76–8; *Female's Friend* (Mar. 1846), 61.

necessities' as they. The implications of a shared nature, as between
prostitutes and respectable women, was carried through remarkably
by some writers. Respectables who had themselves transgressed
were urged to sympathize with the more grossly transgressive, and
those who had not transgressed to remember that they only had
God's providence to thank for their escape: 'English ladies . . .
remember . . . you might have fallen even as they'. Introspection will
confirm it:

> though with all my *enlightened* advantages of baptism, education, confirma-
> tion, ease, and social attractions, I find it difficult . . . to strive against nature
> and inborn corruption in myself, and but too often yield to its evil prompt-
> ings . . . or, from ennui like *pour passer le temps* playing on the edge of
> tempting precipices, until suddenly in some giddy moment, I also fall.[65]

If some of the growing kindliness, as between respectables and prostitutes, did indeed receive an impetus from the fallen-woman fiction
of the 1860s then leniency may have gone far indeed: for the
Anonyma novels, in particular, depict their heroines as wholly normal
and likeable.

But, by one of the strangest paradoxes of magdalenism, the notion
of the moral immunity of the woman worker is also intrinsic to its
operations. The enduring contribution of the early burst of rescue
work, at the beginning of the century, was the involvement of
women: the deliberately 'distinguishing feature' of the London
Female Penitentiary, otherwise modelled on the Magdalen, was 'the
co-operation of intelligent and pious ladies'. As this involvement
shifted from merely sitting on an asylum committee to increasingly
active engagement in prostitutes' lives, women were assured that
moral impurity lay more in *not* taking this alarming plunge: 'excessive delicacy arises in reality from its direct opposite, from a low tone
of moral feeling'.

With the exception of the Female Mission to the Fallen the one
organization which expressly put prostitute rescue into women's
hands was the Anglican Church Penitentiary Association of sisterhood-based homes. Despite Joseph Lyne's demeanour, as described
by Munby, it is clear that the moral spirit of these places was crush-
ingly to assert the purity of the sisters, and conversely the long jour-
ney towards spiritual rehabilitation faced by the few ex-prostitutes

---

[65] *Magdalen's Friend*, 1 (1860–1), 8–9, 86, 202; 'Le Plus Bas' 1859: 165. See also
Interesting Account 1835: 18–20.

they harboured. There was a 'constant spiritual fumigation' in the Penitentiary Association homes, with no danger of any smoke blowing in the opposite direction, from prostitute to sister. Not only was there a taboo on any mention of the girls' pasts (one of the ambiguous persistent themes in penitentiary management, exhibited also at Dickens's Urania Cottage, but in the present instance arising because their sin is 'too horrid' to be referred to), which was sometimes assisted by the conferring of new Christian names, but the penitents behaved with extreme deference to the sisters—who in turn sought to 'show towards the Penitents in their manner, speech, etc, such tenderness and pity as would become a forgiving parent'.[66]

These sisterhoods were the nearest things to nunneries which England had seen for three centuries, and it is perhaps scarcely surprising that they are more obviously anti-sensual in their ideology than any other type of nineteenth-century prostitute rescue organization. Conversely, it was in the traditions of Evangelicalism to see the figure of the prostitute as a token of our common fallen nature, a nature which even the most tireless female missionary must accept that she shared. Also, in the case of the sisterhoods there was generally a big gap socially between rescuer and rescued, which reinforced the attractions of a doctrine of the inviolability of the former: the girls would be inspired by the condescension of these women 'in every possible way so superior to themselves'. It may be that the sponsors of the sisterhoods, in their public utterances, made much of this moral elevation to avoid provoking an already hostile public opinion still further—while privately taking a more magdalenist line.[67]

But if the Evangelical worker was low, the object of her labours was very often, and firmly, lower still. The most outspokenly anti-sensual voice in this camp is Josephine Butler, founder of the social purity movement and leader of the campaign for the repeal of the Contagious Diseases Acts. The nub of Butler's objection to prostitute licensing was that it legitimated an immoral practice, which, in terms of the girls involved, meant that they were insufficiently alive to their own sin. Butler did not mind the stigmatizing that might result from labelling a woman a prostitute, but rather the opposite tendency of the Acts: to make the label seem normal to its wearer. In fact, stigma is to be desired: 'neither are we complaining of the fact that women

---

[66] Nead 1988: 201–2; London Female Penitentiary 1808: 11; *Female's Friend* (Jan. 1846), 3; J. Armstrong 1858: 59, 79; F. Thomas 1992: 184–5; T. T. Carter 1853: 3–4.
[67] Hill 1973: 283; Allchin 1958: 70–1.

whose profession is infamous are kept apart by society. That it should be so is inevitable—is right'. Only in these circumstances will the 'conscience' stay awake. In this sense Butler was not really improving on the existing rescue practices of her home town of Liverpool. In the 1830s the local penitentiary used to display one inmate, Elizabeth Kennens, as an edifying spectacle almost in the fashion of a fairground attraction. This girl, housed in a special room, was paralysed and suffered from fainting fits, supposedly as a result of her 'excesses'. She was thus rescued and yet, very vividly, never free from taint: 'she is as a "brand taken from the fire", and yet remains scorched. She bears in her body the sins of her youth . . . it is in her affliction, her *abiding* affliction, that she is sanctified.'[68]

Confinement: In her views on prostitute rescue Josephine Butler was a good example of the mid-century rejection of the old-fashioned asylums, in favour of more home-like arrangements.[69] This is likely to strike the modern observer as inconsistent with her defence of society's stigmatizing of the prostitute: for the old penitentiaries—which generally admitted girls for a minimum period of two or three years, made them wear uniform (in some instances shaving their heads on entry), and subjected them to an austere regime—look like an excellent instrument of stigmatization. Could it be that the new wave of 'homes' and 'cottages', which made much of the flexibility of the inmate's period of stay, of the relinquishing of uniform, and of the introduction of recreations, concealed an iron fist of humiliation in the new-fangled velvet glove of kindliness?

One may be struck by the sheer persistence of the closed institution, whatever name it bears, as an element in the programme of prostitute rescue. Should its survival be regarded as a hangover from an earlier approach, or as diagnostic of the essential character of nineteenth-century prostitute rescue? There is no question that the closed institution was supplemented increasingly by other lines of attack. That pioneer of a less sectarian handling of prostitute rescue, the Metropolitan Female Asylum, had indicated the main alternative approaches as early as 1829: 'immediate admission' (i.e. day-and-night refuges open to all), protection of those at risk of selling their bodies (this was one of the purposes of offering shelter on demand), and emigration within the British Empire.[70]

[68] Butler 1870; *Interesting Account* 1835: *passim*.　　　　[69] Butler 1928: 45–6.

[70] Metropolitan Female Asylum 1830: 4–5, 9–11, 36–40.

Many of the sectarian organizations, including some of the long-established, did follow suit. The London Female Penitentiary seems to have organized emigration to Australia (admittedly for less than 2 per cent of its inmates) at an early date. By 1860 over a third of the Rescue Society's admissions were preventive cases, and in the first fifty years of its life the London Female Preventive and Reformatory Institution—as was only consistent with its name—handled 26,000 cases of this sort out of a total of 40,000. These girls may sometimes have been submitted to the same regime as actual prostitutes, but the Female Mission to the Fallen, by 1866, only sent just over half of its cases to homes (helping the others straight into work, back to family, and so forth) and regarded 'the preventive cases' as its 'main work'. The proportion of women dealt with by the Mission by techniques other than confinement in fact grew spectacularly at this date. In 1860 every case was handled in a home: fifteen years later only 18 per cent (which amounted to about the same total) went within walls. Perhaps about half the girls contacted by Blackmore and Vanderkiste on their nightly expeditions were given no more than a bed for the night. A quarter of the girls who came forward at the Midnight Meetings were put in touch with employers or family.

Residential provision itself diversified, with the London Female Preventive and Reformatory Institution running a home for unemployed servants and a night refuge (which became by far its biggest operation), while the Female Aid Society from as early as 1836 had made the sheltering of servants a prime purpose. In Bristol in 1859 Frances Power Cobbe founded a mission dedicated entirely to prevention. One has the impression that prostitute rescue was often in these operations simply becoming caught up in the vast wave of philanthropic activity directed at the urban destitute, which in London was coping with literally tens of thousands of cases of homelessness, poverty, and hunger annually in temporary and permanent refuges. It was a natural response of the Evangelically minded to feel that, 'although the prevention of evil is better, and more desirable than its cure, it is the latter that is more open to the kind interference of Christian charity'.

'Cure', and specifically 'cure' under the guidance of Christianity, was evidently the role of the closed institution. And sectarian prostitute rescue clung to this device, even though it was expensive compared with some of the alternative expedients being adopted on a large scale. Cobbe's Bristol experiment is perhaps the only case of a

project, in England at least, specifically targeted at prostitution
which did not have a 'refuge' of some sort in its armoury. The
London Female Mission, for example, from its inception in 1836 had
a wide agenda of methods—suppression, shelter, rescue—designed
for a broad range of London women, and actually always routed a
majority of its cases past the doors of its two asylums. But its never-
forgotten main purpose pointed to these institutions as the heart of
its activity: 'the salvation of the soul is the object of the London
Female Mission'.[71]

Some of the early penitentiaries had regimes which were severe to
the point of cruelty. William Logan (himself an urban missionary
with a less compassionate philosophy than that of Blackmore and his
colleagues) asks penitentiary matrons, in 1843, to consider whether
frequent corporal punishment is really necessary. The unaptly named
Benevolent Society for Reclaiming Unfortunate Females in
Liverpool, one of the cities in which Logan worked, returned its
inmates to ordinary society gradually, so that they would not be
'endangered by the buoyancy of spirit which may follow the sudden
removal from long continued restraint'. While it was certainly much
less the purpose of the new wave of homes to stifle 'buoyancy of
spirit', their supporters had a way of concluding that the old 'bar-
barous' type of home had just carried a good thing too far (and most
penitentiaries in the mid-1860s still had a 'punishment-room', which
subdued by solitary confinement and a bread-and-water diet). The
improved model would mete out a 'gentler and more kindly treatment, but still firm discipline', combining all types of regime 'except
that of force': 'a little of the military . . . a little of the silent . . . a little
of the unrestrained at proper times'. The 'silent' system referred to
here is that of the Anglican sisterhoods, which psychologically at least
were probably the most oppressive of the new homes, not ashamed
of the 'needful severities of discipline', or of 'discipline . . . of a
directly religious . . . kind', 'powerful moral suasion', and 'unswerving
coercion'.[72]

---

[71] R. Knox 1857: 172; *Magdalen's Friend*, 2 (1862), 283; W. J. Taylor 1907: 106; Female
Mission 1866: 7–8, 1867: 34–5, 1868: 31; *Reformatory and Refuge Union Journal*, 46–56
(1870–2), 201; Reformatory and Refuge Union 1877: 17; Blackmore 1860: 21–2; *Statement
1860*: 19–37; *Friend of the Fallen*, 1 (Feb. 1870), 12; Henry Mayhew 1861–2: IV. xxvii–xxviii;
London Female Penitentiary 1842: 10–11; *Female's Advocate*, 4 (1841), 39.

[72] Logan 1843: 45; Bevan 1843: 20; *Magdalen's Friend*, 1 (1860–1), 281; Skene 1865: 7;
Reformatory and Refuge Union Journal, 34–45 (1867–9), 230–4; T. T. Carter 1853: 8;
Morning Chronicle (8 Aug. 1851), 4.

'Discipline', in fact, is the term deployed with a tiresome frequency to characterize the chastening side of the reformed system. It indicates that there will be severities in the regimes of the homes, but severities with an inner, psychological bearing. There was a very general agreement that some kind of internal change in the prostitute's being was the great goal. It was *not* the purpose of the homes just to enable the girls 'to better their position in life'. To this end an element of hardship was in itself salutary, as Josephine Butler argued about the pains of the prostitute's life. In the same spirit, there were laments about the increasing elegance of prostitution in the 1860s, and the consequent blurring of moral distinctions in public places: 'one of the worst features of these places is the facility they thus afford to the fallen to return for a brief space to the presence of the better conducted . . . and to mingle for a time, unrepentant, in that society'. Even Ellice Hopkins, one of the most rhapsodically magdalenist voices of the time, and a great advocate of a more friendly and informal asylum regime, was adamant that the bottom-line penalty of confinement was necessary, 'to exact, if possible, some proof of repentance': 'these poor prodigal children must never be allowed to forget that they have . . . sinned before God and heaven'.[73]

But there were, crucially, great differences in how the goal of inner transformation of the prostitute was interpreted. Ellice Hopkins, despite these remarks, in practice put much less stress on the continuous experience of repentance than did the Anglican sisterhood-based penitentiaries. The latter were the only branch of prostitute work which made their chief complaint about other institutions the failure to concentrate on penance, and accordingly, against the run of the general usage, these Anglican homes retained the term 'penitentiary'. What was looked for in the girls, and what this was supposed to require of the sisters, created an exceptional intensity, on both sides, in the pursuit of 'restoration': that 'constant spiritual fumigation' appropriate to 'a spiritual hospital'. A permanent attachment to the order, for some girls, was in fact envisaged (at Clewer, as members of a subordinate sisterhood of 'Magdalens'), but the roots of this whole approach lay not so much in the high church mentality as in fervent Anglican conservatism. The most traditionalist of all Anglicans of the day, Bishop Henry Philpotts, strongly opposed, 'for her sake', any attempt to bring the seducer of a fallen woman to book.

[73] Charles 1860: 4; *Magdalen's Friend*, 2 (1862), 165; E. Hopkins 1877: 68, 1879: 5.

It must be ensured that there was 'no course left for her' other than 'the sacred, the hallowed rites of repentance'.[74]

Also unmistakably involved in the Anglican programme was religious conversion: it was 'a religious work for CHRIST's sake'. As one would expect, bringing souls into an active condition of religious belief was a major goal for many workers on the Evangelical wing too. There were some who candidly said that there must be 'spiritual regeneration' as well as 'moral reclamation': that the great aim was not 'even to expose the character of sin in all its hideousness and deformity', but rather 'to make each of them a SAVED WOMAN'. But this was surely to multiply distinctions unwarrantably in the eyes of many of the men and women engaged in prostitute rescue, who were seeking for no more than that 'the lost character be restored'. It was at a Midnight Meeting that the 'SAVED WOMAN' was invoked, and these occasions were anomalous in the way they used the psychology of the revival meeting to elicit candidates for rescue (though the missionaries in brothels probably attempted similar emotional effects on a smaller scale). Charismatic religion may have been popular with prostitutes: Ellice Hopkins mentions their liking for Moody and Sankey hymns—and her point is that in the homes inmates should be allowed to adopt any form of worship they like.[75]

If the extremes of high-temperature Anglican monasticism and Evangelical avidity to convert are set aside, there was a broad swathe of prostitute rescue work which was more in harmony with attitudes in the general community than we might imagine. Sectarianism produced an appearance of disagreement which was misleading, as an amusing moment in the columns of *The Times* in 1857 brings out. The editor criticizes the penitentiary system for being based on the assumption that prostitution is 'simply the corruption of human nature, which you cannot remove': rescue workers should recognize that 'we live in a system of causes and effects', requiring 'secular remedies'. But by a nice irony, in the letter announcing the setting-up of the particular project being criticized here, its sponsors had denounced the view that 'there is an evil inherent in human nature, which it were vain to try and eradicate'.[76]

[74] T. T. Carter 1861: 8; J. Armstrong 1858: 9, 59, 76; Bonham 1989: 24; *Hansard*, 75 (1844), col. 878.

[75] T. T. Carter 1861: 8; W. J. Taylor 1907: 111–12; *Magdalen's Friend*, 1 (1860–1), 120, 136; E. Hopkins 1879: 32.

[76] *The Times* (6 May 1857), 7–8.

On reflection, the project of prostitute reformation in closed institutions could never have been based on a pessimism about the chances of altering an individual's nature, but it is also true that these places were usually run by religious sectarians who are supposed to have some sort of belief in the 'corruption of human nature'. Stereotyped antagonisms—of a sort which I have illustrated on a broader front in the first volume of this study—have obscured a common ground of environmentalism, shared by sectarians and non-sectarians alike. It is a mode of thought which, in its nineteenth-century manifestation, is well encapsulated in the analogy between society and the reformatory school which was drawn by the first secretary of the Liberal and philanthropic Social Science Association, George Hastings. According to Hastings, 'the type of the great work before us' is 'a well-run reformatory school', exhibiting 'moral and religious discipline, combined with good sanitary arrangements, and a proper union of industrial and intellectual education'.[77] The Female Mission to the Fallen, one of the great new initiatives in prostitute rescue, was set up in 1858 by the Reformatory and Refuge Union, the body by which Evangelicalism above all demonstrated its commitment to the reformatory principle.

The reformatory school was not a lenient institution. It put its faith in 'discipline', just as the prostitute homes so consistently did. But its severities were not punitive; they were based on an optimistic notion of human capacities, and in that sense on an affirmative attitude towards the inmates. Likewise there is a logic to nineteenth-century magdalenism's combination of a warmly sororal and compassionate rhetoric with certain austerities of institutional theory and practice, a combination which can seem utterly baffling to the modern observer. Finally, the strange moral immunity of the woman rescuer—who is immune even though it is precisely her gender and her gender's sexual susceptibility which qualify her for this work—becomes more intelligible in the context of environmental moralism. This immunity is like the sexual indifference of the gynaecologist, which was claimed by progressive medical opinion but denied by conservatives, or the attenuated sexual appetite of the man of the future, claimed by Godwin but denied by Malthus: it is a token of the plasticity of the libido. There is, quite explicitly, nothing different or transcendent about the sexual constitution of the female rescue worker. But her

[77] Abrams 1968: 49.

background has profoundly altered its functioning, so that the notion
that she could be sexually aroused by involvement in a prostitute's
life is regarded as an old-fashioned libel, worthy of the *Saturday
Review*.

The coincidence of sectarian and secular views on prostitute management is well illustrated, as a matter of literal practice, by
Dickens's Urania Cottage. This was much more the closed and disci-
plined institution of the mid-century type than may be imagined,
physically less austere than the early penitentiaries but psychologically stringent. The girls were not permitted outside its walls except
in accompanied groups. They were allowed to send just one letter—
read by the supervisor—a month, and to receive one parental visit a
month, or a non-parental visit once every two months. (These
arrangements are all virtually standard in prostitute homes.)
Conversation out of earshot of a supervisor was forbidden. The girls
could not make their own beds, in case they secreted something in
them.[78]

Even the physical attractions of Urania (no uniform—though this
had been given up elsewhere too—an element of recreation, and so
forth) had their psychological purpose. 'These unfortunates are to be
tempted to virtue', said Dickens, combining the vocabulary of tough
and tender very much in the manner of those religious workers who
wanted 'a judicious mixture of arguments, appealing by turns to the
animal and spiritual parts of our nature', or who advised that 'you
must bring them tenderly . . . under the yoke of discipline'. The first
step to 'virtue' was to attract the girls, and hold them, inside
Urania—for Dickens attached as much importance to confinement as
any prostitute worker of which we have a record. It was 'the sacred
duty of everyone' on the project '*first to consider how best to get them
there, and how best to keep them there*'.

An important bait was the prospect of emigration, with respectable
marriage; it was in fact the leading note in Dickens's propaganda to
possible inmates. Urania arranged the passage and destination of the
emigrants itself, and this formed such an important and distinctive
part of its operations (in which it was matched only by the old
Metropolitan Female Asylum) that the home has been called 'in reality a colonial training school'. But 'training' was very much to the
point, and only just over half the inmates of Urania ever left England.

[78] *Household Words*, 7 (1853), 170–1.

In this sense successful emigration for a girl at Urania was the last step in a process of selection, and this gave Dickens's project a further important resemblance to the mass of prostitute homes.[79]

Selection: One of the standard complaints about prostitute rescue work was the alleged fact that sexual transgression seemed to be given a special significance. Why, it was objected, were drunkenness or burglary not the subject of such vigorous efforts? To some extent, the objection was simply inaccurate. In so far as the overlapping operations of the law left room, voluntary sectarian effort did go into other kinds of moral rehabilitation. The Reformatory and Refuge Union, which created the Female Mission to the Fallen and helped several other prostitute rescue organizations, put much more energy into reformatory school work, as allowed by the Act of 1854. And it did support a Home for Female Inebriates. At the first sisterhood penitentiary, Clewer, some of the girls had apparently only 'fallen' into thieving or drinking.

Ellice Hopkins actually agreed that it was wrong to make an 'artificial distinction between this and all other sins', and accepted the equivalence with drinking: 'treat her, and act towards her, and feel towards her, as a sister, even as you would treat the drunkard as a brother'. But why, one might ask, did Hopkins not promote homes for alcoholics, in which the latter were treated like prostitutes? And the very terms in which she accepts the analogy with drinking ('treat her . . . as you would treat the drunkard') perhaps denote a fencing-off of a woman's resort to commercial sex as a matter of individual moral pathology, which could prepare for an artificial emphasis on sexual transgression in charitable work.[80]

Increasingly, from about the mid-century, prostitute rescue dealt with women who had never sold their bodies, the 'preventive' cases. But this was still to regard them in a sexual light: as 'unfallen', to use the Rescue Society's blunter classification. One would need to know much more about actual cases than can now be known in order to judge this vocabulary. Many of the women involved may have been quite justifiably regarded as being on the verge of prostitution, either by their own account or through their circumstances. In particular, the 'preventive' cases surely included women who had simply had

---

[79] Dickens 1981: 183; *Magdalen's Friend*, 2 (1862), 251, 286; Dickens 1988: 83; Pope 1978: 186; *Household Words*, 7 (1853), 161.

[80] *Magdalen's Friend*, 2 (1862), 131; Bonham 1989: 20; E. Hopkins 1877: 33.

non-marital liaisons, sometimes with illegitimate offspring as a result. Nevertheless, the Female Aid Society made its prostitute rescue work just one part, about a quarter, of its whole operation on behalf of working-class London women in distress. This system seems to have been unique, and could surely have been imitated with good effect.

Evangelically minded workers did say that the prostitute's sexual activity was an especially serious kind of transgression in itself (and also secondarily, because of its adjuncts of disease and degradation, and the fact of its being a way of life).[81] But this was evidently not a squeamishness about loss of virginity: otherwise the whole magdalenist premiss of rescue work, the reformability of the prostitute, would be nonsensical. In fact a relativistic or qualified moralism was built into the magdalenist enterprise. In the several filtering processes which determined what treatment a prostitute received, the scale of her promiscuity was an important and even venerable criterion. At the Magdalen girls who had been seduced by a promise of marriage were sorted from more hardened transgressors. One London penitentiary, the St Marylebone Female Protection Society, restricted its intake entirely to women who had been 'perfectly well conducted until their fall', and the London Female Mission (though it contemplated at one stage an elaborate system of five specialized homes) would only look at those who had recently turned to prostitution. The tendency of the Anglican sisterhoods was perhaps more towards heroic conquests: Shipmeadow, at least, was 'not a place of refuge for the seduced, but a retreat for open and habitual sinners'. Clewer, interestingly, had up to six places for high-class courtesans. Most asylums operated an age-limit, which ranged from 25 to 30.

In some of these arrangements a widespread assumption about entry into prostitution, that it often started with a loving fall, achieves institutional expression. This was probably believed particularly about the more socially elevated prostitutes, and on occasion a class filter was quite explicitly applied. John Blackmore reserved one of his refuges, the Female Temporary Home, for 'a superior class of woman'; the St James Refuge and Home for Penitents targeted 'women of a superior grade'; and the Church Penitentiary Association did grant the principle that 'it is . . . among the higher classes of these women that the most hopeful material is to be found'. The warden of

---

[81] *Magdalen's Friend*, 2 (1862), 131.

their establishment at Highgate was celebrated for 'the infinite care and pains' he took in vetting cases for admission.[82]

Sheer efficacy, chiefly by the picking out of 'the most hopeful material', was usually the honourable purpose of these procedures. There was sometimes segregation within homes, but more notable was the practically universal application of various thresholds outside and inside the rescue institutions. In the older penitentiaries, girls approved by the committee were passed on to a probationary stage, in an extramural home or probationary ward. The more missionary-minded workers of a later phase used the free-access shelters, or designated private houses, or even their own homes, as a first staging-post. Gladstone was one of those who used his own home, but the Anglican penitentiaries relied generally on urban refuges which they set up as a point for onward clearance to the sisterhoods in rural locations. Shipmeadow, with its policy of tackling confirmed prostitutes, was correspondingly severe in its probationary arrangements: a disused laundry was converted into a first port of call where 'nearly all have to be *broken in*'.[83] Ellice Hopkins put at least some of her cases on a probationary footing for three or four months.

As I have mentioned, the total provision of asylum accommodation did not fall hopelessly short of the prostitute totals: not as far short as the rescue industry—which was always looking for financial support and inclined to accept the highest estimates of the scale of the 'great social evil'—usually claimed. With the massive expansion of non-institutional methods a large fraction of England's prostitutes must have received some kind of help, at some stage, from the rescue organizations. Despite appearances, the latter may have felt fairly satisfied with the numbers passing through their permanent accommodation. Given their goals, the right candidate was in some measure self-selecting: the best guarantee of a reformed life was a strong initial sense of remorse.

Indeed those institutions which placed a great stress on the emotions of penitence and conversion could afford to be in a sense passive in their recruitment. They used the first phase of contact with the missionary to elicit such feelings, which would, if aroused, then carry the girls through to full penitentiary life under their own momentum. This first phase might be a Midnight Meeting, or the

[82] Scudamore 1857: 9; Bonham 1989: 25; W. J. Taylor 1907: 34; *The Times* (6 May 1857), 7; J. Armstrong 1858: 70; F. Thomas 1992: 187.
[83] Scudamore 1857: 9.

period spent in a private house with the missionary, the 'tea and toast' when 'they are dosed with high-pressure religion'. The Female Mission to the Fallen stressed that the great goal, more important than being 'saved from ruin in this world', was that prostitutes be 'brought to the sense of their need of a Saviour': 'the Committee . . . therefore . . . attach great importance to the time spent under the Missionary's roof'. In the early days of the Midnight Meetings prostitutes volunteering to be rescued were taken at once to a home. The religious fervour which impelled them may too often have proved short-lived, or too many turned out to be seeking temporary shelter in hard times (this seems to have been quite common), and after a few years the system was changed, so that women were accepted if they still felt inclined to present themselves in the cold light of the following day. No doubt zeal decayed inside the homes too, but it could be made a requirement for graduating from the probationary level that a girl proved that she had 'evangelical sentiments'.[84]

It may seem that the special systems of prostitute rescue were tailored to the needs of sectarian magdalenists, but, again, they were closely paralleled at Dickens's Urania Cottage. Philip Collins notices that there was 'a change of policy, unnoticed or unrecorded' at Urania, whereby the catchment shifted from prostitutes to any type of female prisoner (Dickens's recruiting was done at Coldbath Fields prison). But this seems to have been forced upon Urania by a shortage of volunteers of the desired sort, even though there were only thirteen places to fill. Dickens's helper at the prison usually had to be content with a 'second-class selection' of girls, who then went forward to Dickens himself for a judgement on 'their frame of mind and general fitness'.[85] Inside Urania there were separate living quarters for probationary inmates. As I have mentioned, the great goal of emigration was not achieved, perhaps not even desired, by many who passed through the whole system.

Dickens's project had the weaknesses of sectarian prostitute rescue—the regimented, threshold-minded, and inevitably stigmatizing confinement in an institution—and none of its strengths. For the sectarian workers, at least on the Evangelical wing, did dispense essential physical support and social restoration among the prostitute population on an impressive scale. They may have preferred to create

---

[84] Judge 1874: 31; Female Mission 1867: 9; *Magdalen's Friend*, 2 (1862), 124; D. Hudson 1974: 82; Davies 1874: 371; London Female Mission 1838: 5.

[85] Collins 1964: 98; *Dickensian*, 57 (1961), 21–2.

passionately remorseful, born-again Christians, but their main achievement was to make many women feel less degraded physically and emotionally, and often to ease them into a non-prostitutional framework of relationships. In fact we have no right to say that this was not entirely satisfying as an achievement for many of these workers, or to insist that their behaviour was not, often, what it appears to be on the surface: a remarkable overcoming of moral, social, and even physical taboos in the interests of compassion.

The magdalenist movement is a wholly nineteenth-century phenomenon, flourishing and then decaying within this period. It is, indeed, the one completely distinctive feature of nineteenth-century sexual culture: more so than masturbation anxiety, or the suppression of women's sexuality, or any of the other elements in the standard portrait of the period. I have argued that its central principle was, like so much in the rest of the culture, progressive and environmentalist, even if it did bring in its wake other specialized spiritual and moral purposes, such as the neo-monasticism of high church Anglicanism. Although Evangelicals were always the most active force in magdalenism, in its various stages, it was not at bottom an Evangelical movement, at least in the sense that traditional Nonconformity was hesitant in its support, and Anglican revivalism, as represented by Wilberforce, preferred a more old-fashioned criminalizing approach to the prostitute.

It could be said, however, that magdalenism was criminalizing enough, for all that. Its roots may have been with the *philosophes* rather than the Methodists, but one token of that background, and the central device of the movement, was the penitentiary—an institution that was always oppressive, sometimes harshly so. In other words, the logical gap between environmentalism and anti-sensualism could have opened up for non-sectarian men and women, as they contemplated the magdalenist operation. The link between the two things is, after all, arbitrary: you may believe that sexual drives can be modified by conditions, but decline to approve of their reduction by this mechanism. The treatment meted out to prostitutes by the rescue organizations, and the spirit and personalities involved, might have made outsiders query the antagonism to sexuality which went along with all this.

The situation is much complicated, admittedly, by the fact that magdalenists did not believe, as a general rule, that prostitutes were motivated by sexual appetite. Nor were these workers physically squeamish or fastidious, in any obvious fashion, about the girls they dealt with: such attitudes were actually ruled out by the magdalenist ethic. But ideas of taint and contamination did get smuggled back into this ethic. It is arguable, indeed, that magdalenism was a way of keeping such notions current, when there was a threat that they would lose ground in a new easygoing sexual culture. There is actually some appearance of a revival of older, pre-environmentalist doctrines of the innately depraved character in the great importance Josephine Butler attaches to stigmatizing the prostitute, as the Contagious Diseases Acts threaten to make the latter feel at ease with her role.

And even if magdalenists were not striking directly at women's sexuality, they were striking at men's. This bring us to a sheer demographic fact which might have been expected to render magdalenism widely uncongenial outside sectarian circles. Prostitutes only existed because they had clients. We do not know how many there were, but they must have amounted to several hundred thousand. They enormously exceeded the numbers of men involved in prostitute rescue work. Even if many resorted to prostitutes only seldom, reluctantly and remorsefully, and would have agreed that prostitution was actually an evil, there must have been many nineteenth-century men also for whom the whole magdalenist discourse and effort was poppycock. After 1857 they had some authoritative backing from William Acton.

'Walter', in *My Secret Life*, is frankly, even determinedly, contemptuous of the magdalenist view of the prostitute: most girls were in the profession, according to him, because they wanted to have orgasms, and they loved their work. No public commentator could have voiced such an opinion, but one may guess that it lurks behind some of the *Saturday Review*'s remarks about prostitute rescue. The *Daily Telegraph* did run an editorial in 1860, condemning the Midnight Meetings, which agreed with Walter on one point: that only ugly and unsuccessful prostitutes came forward to be saved (those 'who have nearly starved at their wretched trade'). But it is a token of the confusion that even classic moralism found itself in on this topic that the *Telegraph*'s first account of a Midnight Meeting only a week earlier had taken a completely opposite tack, finding the behaviour of everyone involved 'highly creditable'.[86]

It is tempting to write off the prostitute rescue movement as a dis-

[86] *My Secret Life* 1966: 745, 2022, 2301; *Daily Telegraph* (10 Feb. 1860), 6, (17 Feb. 1860), 4.

1. William Johnson Fox, London Unitarian minister. Most of his congregation accepted his publicly separating from his wife in 1834, and setting up house with a new companion, Eliza Flower.

2. Edward Irving, a Scottish minister whose teaching of the physical sinfulness of Christ took London by storm in the 1820s.

3. A general view of the Agapemone, the Perfectionist community founded by Henry James Prince at Spaxton, Somerset in the early 1850s. The chapel is the building with the banner.

4. The chapel of the Agapemone. Here, in 1855 or 1856, Prince 'took' a female disciple as a concubine, in a ceremony which remains mysterious.

5. A scene during the last days of London's Judge and Jury show, which —starring its founder, 'Baron' Renton Nicholson - had flourished in the Strand area in the mid-century.

6. An Evangelical ex-Anglican, Revd Baptist Noel, addresses an audience of prostitutes attracted from the night-time pavements of the West End. These 'Midnight Meetings' in London and elsewhere, around 1860, witnessed some of the most personal attempts to reform prostitutes undertaken by the magdalenist movement.

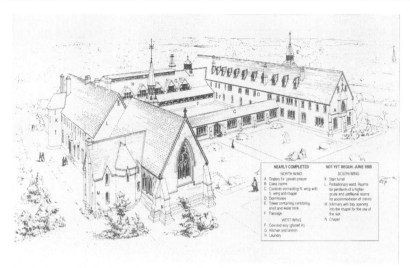

NEARLY COMPLETED	NOT YET BEGUN, JUNE 1855
NORTH WING	SOUTH WING
A Oratory for private prayer	K Stair turret
B Class rooms	L Penitentiary ward. Rooms
C Cloisters connecting N. wing with	for penitents of a higher
S. wing and chapel	grade and additional rooms
D Dormitories	for accommodation of inmates
E Tower containing ventilating	M Infirmary with bay opening
shaft and water tank	into the chapel for the use of
F Passage	the sick
	N Chapel
WEST WING	
F Covered way (glazed in)	
G Kitchen and larders	
H Laundry	

7. The home of the high-Anglican sisterhood at Clewer, Windsor (founded 1849), the first of these bodies to attempt the reform of selected prostitutes (and courtesans) in a convent-like environment.

8. A radical club in Pentonville in 1859 debates issues such as 'full-blown democracy', and refreshes itself convivially.

9. Robert Owen sketched at a Manchester newspaper shop door around 1838, when the Owenite movement was seeking to tone down its bolder policies of sexual reform.

10. Manea Fen, near Wisbech, an ill-starred Owenite community whose free-love arrangements were generally disavowed by the Socialist movement.

11. James Pierrepont Greaves, founder of a vegetarian, pro-chastity community at Ham Common, Surrey, whose doctrines appealed to some leading Owenites.

12. James Elishama 'Shepherd' Smith, whose sexual ideology embraced at different times the ambiguous teachings of Southcottism, Owenism in the early 1830s, and his own 'Universalist' religion.

13. Contraceptive sponges, descendants of devices recommended in the pioneering birth-control propaganda of Francis Place (and later favoured also by George Drysdale and Annie Besant), as advertised commercially in the 1920s and 1930s.

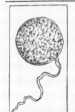

14. George Drysdale. The only known portrait of the reclusive author of *The Elements of Social Science* (1855), which argues the need for unrestrained sexual activity by all biologically mature individuals.

15. Charles Bradlaugh, the century's greatest advocate of free-thought, who founded the first Malthusian League in 1861 in response to George Drysdale's ideas, and stood by the latter's sexual policies despite bitter antagonism from some fellow free-thinkers.

16. Joseph Barker, the fiercest early opponent of *The Elements of Social Science* within Bradlaugh's movement. His religious loyalties ranged at different times all the way from Bradlaughite free-thought to the revivalism of his own north country 'Barkerite' sect.

proportionately noisy operation by a small number of fanatics. Perhaps such a perception explains why historians have never analysed it. I have earlier tried to indicate the extent of secularist magdalenism, as a general attitude, and have cited the many points of resemblance between regimes at Dickens's non-sectarian Urania Cottage and at religious penitentiaries. But Urania was admittedly an isolated venture—perhaps proof that magdalenism could not strike roots in the general non-sectarian (and prostitute-using) culture. It is worth citing the personal experience of prostitution recorded by the well-known late nineteenth-century journalist Robert Blatchford, in his 1931 autobiography *My Eighty Years*, one of the most engaging memoirs of the period. Blatchford did not have religious affiliations, but from the days of his working-class childhood in Yorkshire he seems to have lived by a naturally high sexual code (which he found shared by his young peers: a point relevant to the next chapter). He joined the army, and was based at Aldershot at a time when the Contagious Diseases Acts relating to such areas must still have been in force. 'The girls in the Aldershot dance-halls and taverns worried me', he tells us; 'I wanted to help them and could do nothing'.

Was this the same impulse that magdalenists felt, and simply had the practical means to act on? One of the girls told Blatchford he was like a 'preacher', but the following passage does not read like Josephine Butler, who professed to be concerned for the same class of prostitute. The question of the importance of nineteenth-century practical magdalenism may be said to boil down to whether this passage records the visceral reality of most prostitute rescue work or something quite separate from the latter—a humane response which should have shamed the average professional magdalenist.

I felt for them a kind of wondering respect. They were not wicked. They were friendly. They were loyal, and their queer humility and simple candour of speech were very touching. I heard one of them on a dance-room say to her partner, a young sergeant: 'Do you ever think how we girls are shamed and humbled to give pleasure to you men?'. That speech, spoke so quietly and gently, made the boy turn white . . . They were vitally and awfully womanly, those sullied flowers.[87]

---

[87] Blatchford 1931: 119–21.

Chapter Three

VOLUNTARY RESTRAINT

I

Working-Class Moralism

FRANCIS PLACE, the great London radical, believed that working-class respectability increased dramatically in or around the beginning of the nineteenth century, and he devoted much of his writing and social inquiry to the attempt to prove this claim. To judge from what independent evidence there is, Place's perception was broadly correct.[1] But this is to confirm what is perhaps the less interesting of twin claims which Place made about the matter. He also argued that a great impetus to the new respectability came from the ideals and practice of working-class radicalism. In fact radicalism is the only causative factor cited by Place which can be thought of as ideological (the other mechanisms he finds room for are by way of legal, economic, and educational developments).[2] Religion, in any form, is a particularly striking omission from Place's catalogue of 'good producing causes' in working-class 'manners and morals'. Can this bias possibly be justifiable?

On the face of it, to allow no role for working-class Nonconformity—and especially for Methodism—in these moral shifts is absurd. But as we have seen in relation to middle-class moralism religious affiliation, or the lack of it, may not influence moral codes quite in the manner we foresee. An unexpected logic can assert itself par-

---

[1] See ch. 3 of my *The Making of Victorian Sexuality*.

[2] The most important analysis of influences offered by Place will be found at Place 1972: 14–15.

ticularly where sexuality is concerned, because sexuality has medical, social, and even metaphysical dimensions which mean that it is not on all fours with the other areas of human life—law-abidingness, for example, or use of money, or consumption of stimulants—which the ideal of respectability bears upon. This is equally a problem when the ideal of respectability springs from political roots. In fact in this case there is more of a problem, in the sense that a radical political programme can very naturally find a place for policies on sex which are transgressive by conservative standards: divorce by consent, the acceptance of temporary unions outside marriage, encouragement of birth control, and so forth. The next two chapters will be mainly about the troubled history of such policies within the progressive and secular community in nineteenth-century England.

The moral colouration of Francis Place's political creed, with its emphasis on orderliness, rationality, temperance, decorum, and prudence, may have no necessary connection with radicalism, but it has the strongest possible contingent link with the radicalism of the nineteenth century. One of the persistent motifs in the reports of working-class life written by Mayhew and others for the *Morning Chronicle*—which amount to the fullest account of the subject until Charles Booth—is the distinction between 'respectables' and 'roughs', and an inalienable aspect of being a 'respectable' was a conscious radicalism in politics.[3] It would be true to say, in fact, that there was no segment of the complex movement which composes English working-class radicalism at this date which did not give a central importance to some or all of the Placean virtues.

At a leftward extreme, for example, were the O'Brienist Chartists and post-Chartist groups, whose clubs proliferated notably in London in the 1870s. This was a tradition which consciously resisted an *embourgeoisment* of manners and appearance, but was deeply committed to self-improvement in all non-superficial respects, intellectual and cultural. The London O'Brienists of the 1850s and 1860s gathered chiefly to learn about, and discuss, the most serious topics in politics, science, and literature.[4] These clubs also provided beer, and teetotalism was probably disliked by their membership as much as the 'Teetotal Chartism' of men such as Henry Vincent and William Lovett had been disliked by other Chartist leaders in the 1840s. But the latter example indicates the real affinity that existed between

³ Razzell 1973: p. xxxv.   ⁴ Shipley 1971: 19–27.

ideals of self-control and decorum and the project of political
advance for the working class. Teetotal Chartism may have carried
such ideals too far for the taste of many, but it was a genuine product
of this egalitarian movement, and retained an influence for as long as
Chartism itself survived.[5]

Hostility to drinking had a lively presence elsewhere in the radicalism of the period. Henry Hunt and many of his followers in the
pre-Victorian years were deliberate and influential abstainers. The
Owenites, in various degrees and by various means, opposed drink;
two prominent teetotallers were active in Owenism, and in its early
days the movement sometimes did business in Owenite premises,
even though as a whole it distrusted the materialism of Owenite doctrine. Teetotalism was a distinctively working-class phenomenon,
which transformed the social character of the anti-alcohol movement—hitherto amounting to a middle-class campaign for temperance—as it spread explosively out from its working-class
starting-point in Preston in the 1830s. Though one must not discount
the conservative, non-radical element in the nineteenth-century
work-force, and its contribution to all forms of working-class
respectability, the impetus of the teetotal movement was overwhelmingly a progressive one. At least twenty leading teetotallers were
involved in feminist causes.[6]

By the time the extreme radical clubs of the 1870s were making
their arrangements the mere provision of beer had acquired a sym-
bolic force: generally because of forty years of teetotal agitation, and
in particular because of the teetotal-based Working Men's Club
movement (whose founder, Henry Solly, was a moderate Chartist).
The teetotal belief in the simple efficacy of a dry environment indicates, I believe, that there was much common ground between teetotallers and the Owenite determinism they sometimes made a show of
repudiating.

For many decades the practical congruence between environment
and attitude, in respect of alcohol, was very imperfect. Pubs were
so indispensable and so well-established as sites where the like-
minded or formally associated could gather that even 'a somewhat
embarrassed temperance movement, occasionally met in public-

[5] B. Harrison 1971: 387–9.

[6] *Livesey's Moral Reformer*, NS 1 (1838), 154–5; Prentice 1851: 217–18; Dolby 1827: *passim*; B. Harrison 1971: 173–5, 395, 1982: 213; Faucher 1844: 25.

houses'.[7] There are, inevitably, conflicting accounts of the extent to
which a temperate code maintained itself when working-class radicals assembled on licensed premises. Where the pub accommodation
permitted, some organizations (such as Thomas Wright's trade union
club) were able to meet the ideas of their teetotal members during
formal business—with drinking members repairing to the bars afterwards. George Sala, describing the meeting of a left-wing debating
club in Pentonville in 1859, at which many 'steaming compounds'
were circulating, was 'puzzled to reconcile the full-blown democracy
of some of the speakers, with their mild, bank-account-possessing,
rate-and-tax-paying, housekeeping appearance', and impressed by
the 'sobriety, decorum, and persistence in adhering to the matter in
hand'. The proceedings of a similar club in Birmingham, interestingly
unearthed by Brian Harrison, are most impressive in the variety,
scale, and sophistication of the debates recorded.

   Another observer of such institutions in London judged that even
if the participants were not all 'beery' there was enough of this quality to affect the character of the discussions, and Robert Lowery
recalled that in Newcastle in the mid-1830s the 'warmth of disputa-
tion or the excitement of interesting ideas' actually prompted the
'quickly swallowed draught' rather than 'sluggish sensual sipping'.
We are told that among the Chartists of Pudsey in Yorkshire, who
generally met in pubs, there were many who 'were believers in ale,
and in having a fair share of it'.[8] Two particularly direct attacks on
the drunkenness attending radical discussions describe in one
instance a prosperous, non-activist circle in a London pub, and in
another the 'adjournment to the taverns' after committee-meetings of
radicals working for electoral reform in 1831.[9] It may well be that the
core organizing and planning work of dedicated radicalism was
almost always kept insulated from drink, and that the drunken radical
was generally either an armchair militant or an authentic political
worker consciously relaxing. But even the latter might conceivably
have been a 'believer in ale', that is, a man with a hedonistic code on
alcohol. And it is at the level of general attitude to an appetite and its
satisfaction that the drinking culture of nineteenth-century progressives is relevant to our inquiry. However 'respectable' they were in

   [7] B. Harrison 1971: 52.
   [8] T. Wright 1868: 151–62; Sala 1859: 287–8; B. Harrison 1973; W. E. Adams 1903: ii.
315–17; B. Harrison and Hollis 1979: 82; Lawson 1887: 64.
   [9] C. M. Smith 1853: 251; Burn 1855: 125–6.

other regards, it would be surprising to find that these individuals were anti-sensual in their sexual codes if they had a principled belief in the merits of drink.

This is a complex topic,[10] but it appears that working-class radicals who attacked the various anti-alcohol movements chiefly objected to them on political grounds: for example, that they were motivated by, or assisted, class hostility, or that they interfered with individual liberty. Sometimes, indeed, these objections yield nothing to the anti-alcohol lobby in point of belief in abstinence: Francis Place only resisted the operations of the lobby because he felt they put the working-class in a bad light. Also, the strongly religious colouring of the temperance and teetotal campaigns alienated secularists. The teetotal philosophy in particular, with the abrupt and absolute change of life signalled by the 'pledge', must have irked many individuals not at all opposed to abstinence as a guiding principle.

The great query which hangs over English working-class respectability is whether it was the achievement of middle-class propaganda, for it certainly entailed a convergence with bourgeois standards. Marxist historians, in particular, see the matter in this way. It is a possibility which is obviously very destructive of the picture which respectability-minded working-class radicals, such as Francis Place, had of their own motives. A code of self-discipline, which Place perceived to spring from, and be required by, the growing power of the working class, turns out to be covertly a device for ensuring that the working class defers to the moral and social authority of the bourgeoisie.

In a sense it does not matter, for the present inquiry, whether the Marxist analysis is true or false: there is interest in the fact that Place believed there to be a logical connection between political advance for his class and anti-sensualism, however deceived he may have been about the causes of his belief. On the other hand, one may expect a moral-political system involving this degree of self-deception, and of tension between actual and intended outcomes, to be rather ramshackle, with its various parts less co-ordinated and less illuminating about each other than would be the case with a system autonomously developed. And there is a likelihood that such a system would not be universally supported, for many workers would perceive that it served bourgeois interests better than their own.

[10] Best surveyed by B. Harrison 1982: 142–6.

In the last two decades several important studies have done much to weaken the Marxist explanation of the convergence of working-class and bourgeois ideals in the nineteenth century, and to bolster the proposal that working-class respectability was the spontaneous result of the working-class drive for political advance, or at least (in the words of a historian of class in early-Victorian Leeds) that there was enough of a mix of attitudes, of 'conflict, consensus, deference and defiance', to make 'class-bargaining' the best term for the trans-actions between bourgeoisie and workers. In particular the historian Trygve Tholfsen has argued such a case on an ambitious scale, not shunning the evidence of some extremely bourgeois-sounding utter-ances by working-class leaders, or the extent to which these senti-ments suited the interests of the middle class. He is persuasive that in these men a belief in the traits of 'rationality, civility, morality, self-respect, responsibility' could coexist with an unimpaired sense of the cruel injustice of England's economic and political system, and of the need to change it.[11]

Such a combination of attitudes became increasingly tenable, for after the 1830s the old severities of political economy lost their authority in middle-class circles, and there was a widespread acknowledgement of the legitimacy of working-class aspirations.[12] Politically this whole movement culminates in Gladstone's adminis-trations in the 1880s, and the far-reaching radicalism of the Liberal Party's thinking at this time. It has even been suggested, in relation to the Lancashire cotton-districts, that it is only after the onset of this bourgeois mellowing that working-class respectability feels the sway of middle-class hegemony, and becomes an occasionally divisive force where it had hitherto 'served to build and strengthen class con-sciousness' among the industrial population.[13]

There is what one might call descriptive evidence for the auton-omy of the working-class code of respectability in the accounts given by the French sociologist Le Play and by his follower William Lucas Sargant, a Birmingham industrialist and reformer. According to Sargant, the 'greasy appearance' and 'coarse familiar expressions' of the workers tended to persuade their middle-class superiors that they were beyond moral influence, mere 'degraded specimens of human-ity'—whereas the truth was that the working class had high standards

---

[11] Morris 1990: 119–20; Tholfsen 1976: *passim*, especially chs. 4, 5, 7, 8.
[12] See e.g. Koditschek 1990: 517–20, 551–65; B. Harrison and Hollis 1967.
[13] Kirk 1985: 183–5, 220–30. See also B. Harrison and Hollis 1979: 31.

of personal morality. As I have mentioned, London's latter-day O'Brienists, though they clung proudly to the insignia of their class, were as keen on self-improvement through education as any member of the National Association for the Promotion of Social Science could have wished.

The picture of moral respectability split off, in a working-class context, from its adjuncts in bourgeois culture does create a problem, however: it raises the possibility that the whole logic of working-class moralism may be different from its middle-class equivalent, and in particular that not much can be deduced about sexual codes. Le Play notices, for example, that in some working-class families a very considerable decorum, orderliness, and temperance coexisted with an utter indifference to religion and fecklessness about money. I have in the first part of this study commented on Sargant's very interesting account of factory-hands temporarily transformed at their employer's party.[14]

Unfortunately there is little comment directly on sexual morality in the literature of working-class respectability after Francis Place, or even after Cobbett's *Advice to Young Men* of 1830. Working-class autobiographies tend to be reticent on sexuality (the solitary—and late—exception to this rule is the autobiographical series written by Allen Clarke for the *Liverpool Weekly Post* in 1934–5). This characteristic in itself may be valuable evidence of an autonomous anti-sensualism, for it obtains regardless of whether the autobiographies are unpublished, or published in working-class journals, or published for a middle-class readership. In other words, it is not the result of censorship. By contrast, the essays on aspects of moral reform which members of the working class were invited to write by the National Association for the Promotion of Social Science—with prizes awarded by the firm of Cassell's—were very much for middle-class consumption, and the required moral orientation was plainly given in the subjects sets: 'On Indiscreet Marriages', 'The Paternal Headship', and so forth. Little, perhaps, can be learnt from these offerings about working-class attitudes in general, but the essayists do enlarge enterprisingly on the dangers of the proletarian bathnight for childhood morals, on the need to induce 'remorse and shame' in a fallen daughter, and on other sexual topics. Interestingly, appeals to religion are not conspicuous in the Cassell essays.[15]

[14] Sargant 1857: 301–2, 390; Le Play 1855: 188–9, 194–5, 200–1, 206.
[15] Vincent 1981: 42; *Social Science* 1861: 52, 66–7, 77–9, 143, 196.

There is however one anti-sensual voice in the Victorian years which comments on sexual morality and comes from the heart of working-class aspirations: that of William Lovett, the leader of 'moral force' Chartism. Lovett is actually a more severe moralist on sex than Place ever was, for the latter, though approving of the greater orderliness of working-class sexual codes, was no enemy of sexual pleasure. Lovett can hardly find words strong enough to express his disapproval of non-marital sex: 'in the whole catalogue of vicious indulgences . . . *unchastity* is the most prominent; and at the same time the most insidious, and undermining in its effects'. The physical results of youthful indulgence can be dire, even fatal. Lovett wanted stricter laws on prostitution, and opposed divorce.[16] Though a moderate Chartist in tactics, he was uncompromising in his demands for working-class rights. His involvement in social and moral issues follows on his departure from the political stage, but there is no reason to think that his political beliefs had changed.

Through both phases of his career, and increasingly, Lovett was an enemy of drinking, although he was not among the teetotal Chartists. In his case, certainly, alcohol and sex fell under the same anti-sensual ban. All in all, Lovett's positioning within the radical movement makes it plausible that his strict ideas on sex would have been congenial to many other working-class progressives who are silent on this topic. In a rare glimpse of sexual demeanour in these circles, Ben Brierley recalls how 'on the bursting up of the Chartist movement' his local branch association took up other interests: 'intellectual pursuits' and simple recreations. 'No billiards, no cards, left us time to read . . . We did occasionally indulge in a dance . . . No suggestive hugging was tolerated, and he who ventured upon this objectionable practice was likely as not to get a slap.'[17]

By contrast, it seems likely that John Cleave, who contributed to political militancy and atheism most notably as a publisher and editor of books and newspapers, was a principled sexual libertarian: he is known to have introduced a mistress into his home.[18] Yet Cleave was a teetotal Chartist, in fact one of the four founders of teetotal Chartism. There was at least one argument commonly resorted to by radical advocates of temperance and teetotalism which had nothing to do with drinking as an appetite or pleasure, namely, that working-class drinking suited the interests of the ruling classes and was

---

[16] Lovett 1853: 77–83.   [17] Brierley 1887: 49–50.
[18] Miles 1988: 237–8.

encouraged by them accordingly. This is a point at which the logic of working-class respectability certainly differed from its bourgeois counterpart. It is necessary to ask, to what extent were the underlying assumptions of this respectability such that a code of sexual reticence and restraint were likely to form part of it?

On the whole, and despite the case of John Cleave, it is probable that antagonism to drink, for example, and a strict sexual code would have gone hand in hand in the thinking of working-class radicals, precisely because of the form of their radical creed. Lovett offers one line of thought linking sexual discipline and political progress which springs easily from his belief in the dangers of sexual indulgence, but which may not have recommended itself to other radicals: he argues that wealth and power are mainly objectionable for the suffering they cause, and thus can be bracketed with sexual licence as 'animal propensities' which must be condemned on utilitarian grounds. But Lovett's other line of thought on sexual morality would have found an echo in many minds, being a very typical statement of the environmental moralism so universal in all progressive or quasi-progressive circles: 'man . . . has within him the capacities of the philosopher and the propensities of the savage; and whether he shall be one or the other will depend on the . . . means taken to develope the good and control the evil'. As with so many of his contemporaries, for Lovett education is the key environmental 'means' to employ: 'sound and wholesome knowledge' about sex imparted to the young will build a 'moral rampart in their minds'.[19]

There is evidence here, to start with, of an implicit superiority attributed to the 'philosopher' in human nature as opposed to the 'savage' and the 'animal', which tends to discredit all physical appetites and satisfactions alike. Obviously, this is in a tradition at least as old as Plato, and in that sense hardly remarkable. But the Platonic value system was reinvigorated in the Enlightenment, particularly by eighteenth-century radicalism's trust in the intellect and its tools—in unhampered inquiry and expression of opinion—as the means of political progress. In the work of some Enlightenment thinkers and their English heirs, in fact, a high valuation of rationality, and of an associated anti-sensualism, is carried to the point where, flying in the face of the more customary talk of 'savage' and 'animal', it is these qualities which are identified as primitive and original in humanity.

[19] Lovett 1853: 4, 10–12, 80.

The French *philosophe* Volney, in a short treatise published in cheap form in England as *The Law of Nature*, drew from the lessons of 'nature' a moral scheme which E. P. Thompson has likened to 'a prospectus for the Victorian age'. Among the natural 'private' virtues are temperance (which includes sexual continence, physical modesty, and so forth), and among the 'domestic' conjugal fidelity. In England the bold working-class political theorist John Francis Bray expressed the naturalness of human rationality in a way that courts paradox: 'man now exists in an unnatural state—as an instinctive rather than a rational being'. We will encounter similar arresting inversions, specifically in relation to animality and sexual continence, in Robert Owen's thought. Rousseau, as I have mentioned in Chapter 1, offers in *Emile* the view that sexual drives in the male before the age of 20 are the artificial product of a civilized environment, which the Owenite freethinker Charles Southwell hailed as an important truth.[20]

The radical tradition puts a great stress on individual rights and individual capacities, such that the individual not only can and should be educated, but may also be his or her own teacher. Self-education, for the qualities of self-control, orderliness, and aspiration which it brings in its wake, is another important link between progressive environmental moralism and a broad anti-sensualism. Self-education is in turn part of the larger nineteenth-century ideal of self-help: an ideal which has a well-established intimacy with respectability, and which has been tossed by the same controversies among historians. It is indeed one of the most challenging points of convergence between working-class and bourgeois moral ideology, highly equivocal as evidence either for the autonomy of working-class respectability or for the hegemony of middle-class values.

The most famous sponsor of the self-help philosophy, Samuel Smiles, is himself an equivocal figure. In his thirties, living in Leeds, Smiles was a middle-class radical, a supporter of many progressive causes, including Chartism. The self-help creed was first enunciated by him at this time, in the mid-1840s, in lectures to working-class audiences for whom he seems to have had a true fellow-feeling. Despite his determined hostility to Owenite and other socialism, and his later unpopularity with many radicals, it is arguable that Smilesian self-help always bore the marks of its origins in radicalism. It was a philosophy for the individual, not the mass, to practise,

[20] E. P. Thompson 1968: 815; Volney ?1846: 9; Bray 1839: 40; Southwell 1845: 19.

Smiles insisted, but its goal was nevertheless the raising of the mass, of 'the entire condition of the working man', through myriad individual advances.

Asa Briggs has noted in Smiles's writing a typical example of the period's phantom warfare on environmentalism. Smiles affected to deplore the Owenite 'axiom that character is formed *for*, not *by*, the individual; and that society may so arrange "circumstances" as to produce whatever character it pleases'; but he was a great admirer of organized discipline, and of the power of military drill, for example, to transform the character of working-class recruits. In the 1840s he declared his belief in the factory as an environment which could determine the morals of its operatives.[21]

Smilesian self-help also shows how such an ideal, even as voiced by a middle-class Victorian, could thrive quite independently of religion. The evidence of Smiles's life and work is that religion was of little or no real importance to him: it certainly plays no essential role in the case for self-help. I have commented on Francis Place's resounding silence on religion as an agent of working-class respectability, an implied denial which becomes quite explicit almost a hundred years later in Will Thorne's remarks about the rough morals of the Birmingham gasworkers in the 1870s:

This drinking and fighting should not be magnified or misjudged. We were healthy, normal human beings . . . we had little amusement and little opportunity to enjoy the better things of life . . . We were poor, ignorant victims of the growing machine of industrialism, and the progress we have made since those days has not been because of the Christian hearts of the more fortunate ones above us, but because of our innate love of freedom that has been developed in the incidental battles for bread.[22]

Thorne, a trade union leader and then one of the first Labour MPs, spent much of his life in occupations where the old working-class unrespectability lingered most obstinately: as a navvy, for instance. His comments here are characteristically loyal and uncensorious about working-class roughness, while environmentalism easily copes with the appearance of depravity. Also striking is the confidence that increased respectability (which is not denied to be 'progress') grows straight out of radicalism, out of the 'love of freedom' as intensified in political struggle.

[21] Briggs 1965: 124–47; Smiles 1846; Tholfsen 1976: 127–9, 205–8; Tyrrell 1970.
[22] Thorne 1925: 45–6.

It does not, according to Thorne, owe anything to religion, and he denies this in such a way as to imply that the true proletariat of the day was irreligious. If Christianity had played a role in working-class respectability, its origin, according to Thorne, would have been 'the more fortunate ones above us'. The nature of the connection between working-class religious affiliation and working-class radicalism has been the subject of a resourceful, even brilliant debate among historians.[23] Happily, there is no need here to take sides in this controversy, since, regardless of whether religion tended to promote or inhibit radicalism, it is clear that a strong circumstantial link obtained between Nonconformity and political militancy. Will Thorne is misleading: there were many left-wingers who were also chapel-goers and even preachers.

Progressives across the spectrum from Smiles to Thorne *could* value moral respectability without invoking, or even while repudiating, religion, but it is plausible that religious allegiance, where it existed, bolstered the respectabilist creed of radicals. Thomas Hardy, founder of the London Corresponding Society, seems to have taken commitment to a 'sedate and serious' code rather further than his colleague Francis Place; perhaps the extra edge of anti-sensualism can be attributed to his devout Nonconformity. The contribution of religion as an agent of moralism could be expected to be relatively stronger with working-class individuals whose political commitment was not strongly developed. In some rural areas there may, late in the century, still have been no concept of a secularist respectability: it was a choice of pub *or* church.[24] We need to assess, in fact, the extent of religious belief in the working class as a whole—especially Nonconformist belief—and its likely impact in the moral domain.

The most powerful single determinant of levels of church and chapel attendance in the nineteenth century turns out, surprisingly, to be geographical region. Almost as important is economic status, with the levels in question consistently correlating inversely with the rates of child mortality in different districts. The distinction of urban and rural is the third influence on formal religious activity. On Easter Sunday 1851 about 13 per cent of the population of England's main towns and cities, and about 22 per cent of the rural and small town population, were counted as attending Nonconformist services. Not all these worshippers were of working-class status (and even in

---

[23] For a survey see H. McLeod 1984: 49–56.
[24] T. Hardy 1832: 4–5; Snell 1936: 33.

industrial areas there was an element of working-class attendance at Anglican services—perhaps gaining strength from the mid-century onwards precisely because of the extent to which Nonconformity was the religion of the bosses). On the highest possible estimate less than a third of workers and their families outside the biggest urban centres went to the chapel, and less than a quarter inside these centres.

These may seem like substantial, though not impressive, fractions, but their actual composition meant that there were very broad areas of formal irreligion in the nineteenth-century working class. The correlation between status and religious activity had a strong effect at the bottom of the social scale. The registers of births, deaths, marriages, and baptisms in the most substantial and vigorous of the Nonconformist sects (the several varieties of Methodism, the Congregationalists, and the Baptists) suggest that skilled workers and craftsmen and their families were greatly overrepresented in these congregations. When these sects were at their most expansive they may in fact have recruited from a broad range of the working class, with the superior character of the resulting membership being helped by upward social mobility plus the Nonconformist 'lift', as it has been called, of members' behaviour and aspirations. But it seems that the leading Nonconformist sects all experienced a downturn in fortunes around 1840, and certainly by 1850, and thereafter lost ground in the sense that they grew slower than the general population. In particular, as the 1851 figures would suggest, they probably failed to make headway with the new industrial working class of the larger cities.[25]

These patterns are all deduced from the records of fairly committed adherence to Evangelical Nonconformity, however, and there was among the English working class certainly a much larger following for Nonconformist Christianity which had no specific denominational allegiance. It is uncertain how much the history of the core of committed chapel-goers says about the larger dissenting constituency. The post-1850 decline in recruitment presumably indicates a smaller pool of individuals ripe for the transition from vague Nonconformist belief to denominational commitment: and the chapels' failure to recruit was greater than the sheer figures of affiliation suggest, since after 1840 there was probably less leakage through the lapsing of converts than had been the case.

[25] C. G. Brown 1988; H. McLeod 1973, 1974: 31–2, 293, 1984: 32; Kirk 1985: 205–6; Gilbert 1976: 30–6, 60–7, 146–8, 159–60; Malmgreen 1985: ch. 5; Inglis 1963: 8.

Was the social composition of non-specific dissent different from that of the formal variety? For example, servants were grossly under-represented in the figures of chapel affiliation, and, it has been said, 'provided almost no Evangelical Nonconformist members or adherents' in the early nineteenth century, though they composed about 10 per cent of the population. To judge from the 1837 diary of the London footman William Tayler, servants would as a matter of course attend a variety of Nonconformist places of worship (in Tayler's case a visit to a Roman Catholic church was also thrown in, 'for a fancy', among the diet of Nonconformity). Industrial working-class families may have been committed Bible-readers although not attached to any chapel, and have attended services with more regularity, at a variety of chapels, than the statistical record has been able to detect. Thomas Carter, who worked as a journeyman tailor in the early years of the century, stuck out like a sore thumb among his workshop colleagues for his clergyman-like demeanour and principles—yet he was not exclusively attached to any one denomination. Working-class autobiographies of the period in fact often record such 'theological travels' around different places of worship. When George Lansbury was a child in Bethnal Green in the 1860s his mother operated on the principle that 'for ordinary life the Church of England was enough', but that a visit to the local Primitive Methodist chapel was required if her children had been particularly naughty, and, once, when they had a brush with atheist propaganda.[26]

The fact is that religious belief seems to have been very tenacious in the nineteenth-century working class, though often in forms which puzzled and annoyed orthodox sectarians, and may still be found puzzling. One London missionary records in his diary in the early 1860s how he found a man in one of London's poorest areas, St Giles, who was admittedly 'very well versed in scripture' but completely indifferent to the missionary's approaches. Edward Clodd, a London clerk in the 1850s, went on attending a variety of churches, for the sermons, well after he had lost his youthful Baptist faith. In general Nonconformist missionaries did not complain that they were up against sheer atheism. Some described in their published literature how they encountered very overt boredom or derision in the slums; one need not accept the stories of heroic conversion work for which

[26] Gilbert 1976: 67, 179–81; Tayler 1962: 59; H. McLeod 1984: 14; C. Shaw 1977: 192–3; Ainsworth 1977; T. Carter 1845: 129–30; Hollingshead 1895: i. 43–4; G. Mitchell 1968: 78; Lansbury 1928: 28–9.

these admissions are the flattering frame, but Mayhew learnt that there were some individuals who prayed at their bedsides in the common lodging houses, although tracts left by missionaries at these places were laughed at and used to light pipes.[27]

Even where they did encounter blatant atheism the missionaries need not have despaired, for nineteenth-century atheism was curiously adjacent to religion. To start with, there was an important agnostic wing to the secularist movement, represented most notably by G. J. Holyoake, which was scrupulously non-committal on ultimate questions, and even tended to speak up for religion against atheism proper. To this grouping belonged the Unitarian-descended free-thought of South Place, which the Methodist George Howell was happy to include among his ports-of-call as a worshipper in the mid-century. Secondly, even out-and-out atheists had a remarkable way of reversing religious polarity. The outstanding example is Joseph Barker, who was at different times (and among other things) a charismatic Methodist leader and a Bradlaughite free-thinker—and 'cocksure after every turnabout'. A similar psychology is revealed in the autobiographies of two individuals of working-class background who were not nearly as conspicuous as Barker, the bookseller James Lackington (whose story only just belongs in the nineteenth century) and the Coventry artisan Joseph Gutteridge.[28]

To argue effectively against the detail of Christianity, and even to wish to do this, probably requires a background of Christian upbringing which may reassert its power at any time. A more glib and less persuasive proposal is that these latter-day St Pauls were acknowledging God in the mere fact of their hostility to him. But there may have been something about the spirit of a sweeping atheism which had a psychological affinity with a sweeping supernaturalism, especially in its dissenting version. Ben Tillett, an early Labour MP, says that his 'mysticism' kept him from being converted to the atheism of Charles Bradlaugh and Annie Besant, but that this was the only other creed which could have gained his allegiance. A Northumbrian miner of the same generation, who did entirely repudiate the strict Methodist faith of his childhood, felt that in thus 'demanding the right of free inquiry I was but extending the Nonconformist mandate

[27] J. Kent 1973; H. McLeod 1974: 52; Clodd 1916: 4–5, 10; Inglis 1963: 115; Vanderkiste 1852: 275; Henry Mayhew 1861–2: i. 252, 412. See also Lewis 1986: 125–8, 133–48.
[28] Leventhal 1971: 20; W. E. Adams 1903: ii. 400; Lackington 1791: *passim*, 1804: *passim*; Gutteridge 1969: *passim*.

of freedom of conscience to its logical issue'—rather like the working-man who is reported to have cried out in a meeting addressed by the liberal but high church Stuart Headlam 'atheist as I am, no man shall stand between my soul and my God!' Annie Besant's atheism, as is well known, mutated back into theosophy. At an earlier period, Richard Carlile's successive doctrinal positions offer a baffling convergence of rationalism and certain crudely superstitious elements.[29]

Harder to assess are the implications of this pervasive working-class religiosity, or quasi-religiosity, for moral belief. The St Giles man 'well versed in scripture' told his missionary visitor that 'do to others as you would be done by' was the extent of the moral doctrine enjoined by religion, and Christianity and its institutions were probably respected by many working-class men and women as the foundation of Golden Rule ethics, but no further. They were often incredulous that God would punish violations of this code in the afterlife, let alone offences against more moralistic prohibitions.[30]

Christian forms could be embraced by the working population when they were perceived to favour the latter's purely secular interests. The most obvious instance is working-class Sabbatarianism. A Sunday as active as an ordinary weekday was, presumably, a Sunday on which working-men would have to labour, perhaps for less money, and there was accordingly a degree of radical support for Sunday closing. For what it is worth, a competition for working-class essays on the sabbath, in 1849, attracted 1,045 entries, but one which set 'Man's Duty to God' as a subject, in 1858, received just 54 submissions. Sunday schools were also resorted to enthusiastically by working-class parents, but this was part of a pattern of considerable discrimination in their approach to schooling, in which the educational advance of their children was the overriding concern.[31]

William Tayler, the London footman, had more than a Golden Rule moralism. He found the *décolletage* of the young ladies he attended 'disgusting': 'they are nearly naked to the waist . . . the breasts are quite exposed except a little bit coming up to hide the nipples'. Though Tayler was also very interested in religion there is no sign that these censorious moral reactions were particularly connected with what he felt on his various forays to chapels and

[29] Tillett 1931: 77–8; C. Armstrong 1938: 101; Budd 1977: 45; Wiener 1976.

[30] H. McLeod 1974: 51–2, 1984: 63; Ainsworth 1977.

[31] B. Harrison 1982: 132–5; Bickersteth 1849: *passim*; Rochester 1858: *passim*; Laqueur 1976.

meeting-houses. However, Thomas Carter, as I have mentioned earlier, had strict moral ideas which were associated with powerful religious convictions, even though the latter were denominationally
unspecific. East-Ender Thomas Okey was brought up in 'a religious
atmosphere . . . of the narrowest Victorian evangelicalism', but 'none
of our elders . . . attended service'.[32]

The Chartist Thomas Cooper, who was 'strict' in his moral views
while still a free-thinker, did in due course convert to a belief in
God's existence out of an imperative sense that the latter was
required to give the Christian moral code its deserved authority:

I had taught morals, and taught them strictly; but the questioning within,
that would arise, day by day, and hour by hour, made my heart ache. 'Why
should man be moral? . . . why have you not taught your fellow-men that
they are answerable to the Divine Moral Governor, and must appear before
Him . . . ?'

It was not a conviction of the truth of Christianity, of the reality of the
Miracles and Resurrection, or of the Divinity of Christ, that had worked the
change in me. I was overwhelmed with a sense of guilt in having omitted to
teach the right foundation of morals.[33]

Religious conversion on such narrowly moral, indeed moral-
philosophical, grounds was probably rare, but there must have been
an element of commitment to the moralistic involved in many com-
mitments to the chapel—perhaps formalizing an existing strictness of
standards such as came naturally to William Tayler. What kind of
moral culture did the convert encounter? More specifically, how central was anti-sensualism to the practice of the main Nonconformist
denominations?

The contemporary evidence on this subject is beset by stereotypes,
but the stereotypical accusations of Nonconformist moral laxity are
much less convincing than the images of an energetic and strict
restraint of behaviour and demeanour practised, and taught, by committed dissenters. There is after all the brute fact that the leading
wing of Nonconformity, the Methodist churches, was wedded to
sobriety in dress and utterance, and opposed to most kinds of recreation: for chapel membership to be granted, and then continued, the
moral standards of families were sometimes actively investigated.
Allegations that Methodists were sensual in their standards always
had to go more or less wide to outflank this plain truth. There is

[32] Tayler 1962: 36; Okey 1930: 149.   [33] Cooper 1971: 352.

necessarily a diffuseness and indirectness to these attacks, whether they are relatively enterprising—as in the case of Leigh Hunt's attacks on the sensuality of Methodist liturgy and devotional rhetoric, and of Robert Fellowes's would-be statistical proof of Methodism's depraving effect on national morals—or more traditional: portraits of the dissenting minister with 'jocund visage, and not too narrow in the shoulders', and alarms about antinomian licence to sin. The terse but strangely moving diaries of William Swan and his son, committed London chapel-goers who were isolated among otherwise non-religious kin, contain fascinating records of the interplay between sexual wishes and anti-sensual religious precept. There is a fairly transparent squaring of one by the other, but also a degree of sexual continence apparently authentically sustained by the men's faith.[34]

Two qualifications about Nonconformist moralism may still be entered, however. There is evidence that as Evangelical dissent lost its expansive thrust, around the mid-century, and became more contentedly bourgeois, its moralism changed in character, if not in weight, becoming less singular, militant, and punctilious. And all along there was at least a potential in dissenting codes for some aspects of sexuality to be exempt from repression. I have illustrated in the first part of this study, in relation to Malthus, how there was a general inclination at the low church and Nonconformist end of the religious spectrum to distrust the restraint of licit heterosexuality as an interference with God's purposes. This distrust sometimes produced distinctly pro-sensual accounts of young marital sexuality. One Nonconformist autobiography, that of the Nottingham-born cabinet-maker James Hopkinson, shows such a cluster of values being lived out. Hopkinson depicts himself as the ascetic and highly self-disciplined product of a strict Baptist upbringing, but unlike the Swans is also unembarrassed about the urgency of his wish to marry and about the excitements of courtship and young marriage.[35]

This strand in the Nonconformist ethos possibly moved in a contrary direction, however, to the moralism which concerned the externals of behaviour. While the latter became more lenient in the course of the Victorian period there appears to be a retreat, at a rather earlier date, from the boldest dissenting positions on God-given

<hr />

[34] Tholfsen 1976: 66–7; 'On the Evangelical Sects', *Quarterly Review*, 4 (1810), 480–514; Hunt 1809: 56–8, 90–5; Fellowes 1801: 16–19; *Metropolitan*, 4 (1832), 399; Swan 1970: 5–6, 17–20, 70.

[35] Gilbert 1976: 158–62; Lawson 1887: 73; Hopkinson 1968: 71–96.

sexuality—a development which echoes the toning down of the more
startling features of Swedenborg's teaching. An interesting line can
be traced from Martin Madan in the 1780s to Benjamin Parsons in
the 1840s, both members of the branch of Methodism known as Lady
Huntingdon's Connexion. Madan, who himself was influenced by
Methodist predecessors such as Westley Hall, brother-in-law of
Wesley, is a friend of marital sexuality in a most unusual sense, in
that he wants the conditions of marriage to apply to *all* liaisons based
on sexual intercourse: thus the female sexual partners who at present
are forced into the role of prostitute would be on an equal footing
with the wife, indeed all would together form the wifehood of a
polygamous family. Though Madan was chaplain at the Lock
Hospital his remarkable ideas seem to spring more from his faith
than from his dealings with prostitutes: the *Thelyphthora* of 1780
(enlarged in 1781) is virtually a work of biblical commentary, in
which an enormous weight of scriptural evidence, from both testa-
ments, is adduced in support of polygamy.

There is actually more celebration of physical pleasure in Parsons
than in Madan, but also much more belief in the corralling of sexual
drives within ordinary and orderly matrimony—to be achieved
where necessary by the mind's control over the libido. Parsons
believed that the physical side of our being is undervalued, especially
by 'many good people', and poorly understood (his views here were
similar to those of his colleague in the temperance movement, J. E.
Ritchie, and an example of how attitudes on sex and on drinking
could diverge). In this affirmative vision of our carnality there are
elements deriving from the current physiological doctrine of the
greater sensuality of women, and perhaps from the Swedenborgian
belief in the sexuality of the resurrected body. However great the
pleasures of marital sexuality, however, Parsons believed that their
indulgence should and could be postponed by young people, and in
this sense departs from Nonconformist orthodoxy on the population
question. While moral restraint is being exercised 'intellectual' intercourse, conversation, may fill the vacuum. Parsons can sound confusingly like Rousseau in his trust in the controllability and hence
artificiality of sexual drives: 'take away thought from the human soul
and our appetites would never be kindled into passions'.[36]

It is arguable that the true line of descent from Madan's

---

[36] B. Parsons 1842: 43–5, 234; Ritchie 1857: 24; Hood 1856: 87–8.

polygamous ideal, in sectarian terms, lay not within the church he himself was attached to, but in the unofficial but sometimes very large popular religious movements such as Southcottism. These represent a third mode of working-class religious experience, in addition to unfocused Nonconformity, and deliberate commitment to one of the recognized non-Anglican denominations. Most importantly, for our purposes, there is generally a much more frequent outcropping of specifically sexual concerns in the life and doctrine of these movements than in orthodox Nonconformity. Not that the tendency of these outcroppings is necessarily transgressive and libertarian, however. In fact an ambivalence as between this pole and an opposite extreme, of sexual austerity and even self-mortification, is something of a rule.

I have already mentioned the career of one of Joanna Southcott's heirs, George Turner. John Wroe, whose following was especially strong in Ashton-under-Lyne and Bradford, was the next dominant leader in the disputed Southcottian succession after Turner. Circumcision, which he publicly underwent himself in 1824, was a key feature of Wroe's Judaistic brand of revivalism. The whole mode of his sect was such that it attracted accusations of sexual licence even though, in the early period of its activity, Wroe may genuinely have practised an austere sexual code. But, as so commonly in this sectarian climate, austerity and licence were closely allied. Some years later Wroe's objective transgressions—which involved some kind of flagrant concubinage, perhaps with one particular woman 'marked out' as destined to bring forth a Messiah with Wroe's help—wrecked his standing in the movement in England, and he turned his attention overseas, becoming especially active in Australia. Except for the failure to hold his followers together, the episode resembles events at Henry James Prince's Agapemone a quarter of a century later. Also like Prince, Wroe in his subsequent career taught a message of millennial abstinence: for the last thousand years of the world's history the chosen people were to propagate by virgin births. At some point the Wroeites invented a bizarre ritual called the 'Law of Christ' which well expressed their ambivalence about sex. It was notionally a form of punishment for male Wroeites, but took the form of a spanking by a naked woman who held the victim's genitals, and perhaps masturbated him.[37]

---

[37] J. F. C. Harrison 1979: 138–52; Fielden 1856: 8–15; A. Stewart 1864: 5–7.

Into the vacuum created by Wroe's fall from authority stepped John 'Zion' Ward, who had become a Southcottian after a complex history of other sectarian affiliations (including Sandemanianism). As his sobriquet implies, his revivalism was also modelled on the Old Testament, but castration (of a figurative sort) features more in his teaching than circumcision. In fact Ward seems to have been consistent in his theoretical and practical advocacy of maximum, if not complete, sexual self-denial as a requisite for salvation. He is only ambivalent in the sense of teaching that matrimonial sex, when indulged in, may have a Swedenborgian fineness to it (thus dismaying at least one unmarried female follower), and simply in the way sex looms large in his thought. Here he resembles two of his converts who later came to have links with Owenism: James Pierrepont Greaves and (more significantly) James 'Shepherd' Smith. I shall comment later on Smith's louche toying with themes of sexual abstention.[38]

Why does sex press forward so persistently in the history of these popular religious movements? One reason must be that the whole spirit of such episodes was disinhibiting: this was religion based on fervour and newness—on spontaneous, excited, and even transported reaction to charismatic leaders who were breaking bounds. Both in its charismatic and in its revivalist aspects, popular heterodox Christianity expressed important tendencies of working-class religious feeling generally. Urban missionaries found that ecstatic experiences were part of the unfocused creeds which they were trying to supplant, and even the pattern of recruitment to the recognized denominations, especially in the first half of the century, is notably spasmodic, a matter of revival followed by slump, followed by revival again.[39]

And the element of newness and bound-breaking often went, literally, beyond religion into the whole human condition, social and political. In so far as popular revivalism trenched on visions of social and political transformation it was making contact with another deep feature of the thought of the working class, namely its aspirations for secular advance. In fact in the area of sexuality it is not really possible to speak of strictly secular reformism, at least in the first half of the century. If those popular religious movements which incidentally offered new teachings on sex tended to have a political colouring it is

[38] J. F. C. Harrison 1979: 152–60; Holinsworth 1899–1901: iv. 146–8, 154–9, vii. 108–12, 119, 136–64, x. 89.
[39] H. Macleod 1984: 63; Gilbert 1976: 192–3.

even more the case that new programmes for sexual relations put forward at this period were bound up with religion, in however peculiar a guise.

II

Owen and Company

In the late 1830s, in the years around Victoria's accession, there was some prospect of a sexual revolution in England. That is to say, the most dynamic working-class movement of the day, Owenite socialism, had incorporated into its programme policies on heterosexual unions which to a greater or lesser extent were subversive of existing codes on marriage, divorce, and the respective rights of sexual partners.

Frustratingly, it is not apparent exactly how much support these novel policies commanded. Even if they had been endorsed by all the men and women who counted themselves as Owenites at this period (and it emerged that they were not so endorsed) it is unclear what this would have meant numerically. Estimates of the readership of the *New Moral World*, the weekly organ of the movement in these years, have been pitched as high as 400,000, and estimates of the numbers attending socialist lectures on a Sunday in the 'Halls of Science' as high as 50,000 (with many would-be attenders being turned away for lack of space). But it is also possible to arrive at plausible calculations much lower than this.[40] Obviously, the hard core of Owenism was much smaller than, say, the core of committed Methodists of all types at the same period, but the members of more vaguely loyal socialists may have matched the latter figure.

So a determined effort to change sexual codes, prosecuted by the full Owenite movement, would have been no hole-and-corner affair. And the dissenters from Owenite sexual policy within the movement were balanced, if only to a small extent, by men and women who approved of this particular aspect of socialist thinking, while not otherwise holding Owenite views. W. J. Fox's wing of the Unitarian church was well-disposed, as was John Stuart Mill, who declared in his autobiography that 'in proclaiming the perfect equality of men and women, and an entirely new order of things in regard to their

---

[40] J. F. C. Harrison 1969: 217; B. Taylor 1983: 120, 300; Podmore 1906: 469.

relations with one another' Saint-Simonianism, Fourierism, and the Owenites had all performed a valuable service.[41]

But Mill and Fox, as we have seen, had divergent liberationist views on sex, and where the three movements praised by Mill stood in practice on sexuality was not always clear—especially in the case of the Owenites. From the utterances of Owen and his disciples in a four- or five-year period around 1838 it is possible to derive a code which is boldly opposed to monogamy, but also one which seeks nothing more than provision for divorce when a marriage breaks down—and one which is hedonistic in spirit, but also one which is anti-sensual and high-minded on physical sex. The whole Owenite episode meshes, in fact, fascinatingly with attitudes at large in nine-teenth-century England. Above all, it offers a uniquely stark conjunction of environmentalist thought and the theory of sexual morals, and thus has a special importance for the argument of this book.

The one unshakeable article of faith in the Owenite creed—which otherwise tended to exhibit remarkable changes of direction—was its out-and-out determinism, or insistence that men and women are entirely the product of their circumstances. The Owenite literature can sound crankily emphatic on this point, but Owenism contributed more to the mainstream environmentalism of the period than may be supposed. One historian has seen it as the meeting-point of secular environmentalism and the Evangelical spirit: I have argued in the first part of this study that the latter, in aspects of its philanthropy and of its conversion work, gave a dominant moral role to environment. Looking back in 1880, the one-time Owenite Thomas Frost made claims for the impact of Owen's determinism which sound extraordinary but which are actually a sober summary of facts. According to Frost, socialism's 'fundamental tenet, "man is the creature of circumstances", may be recognized in all the legislation of the last quarter of a century', though the architects of this legislation, men like Lord Shaftesbury, have often been far from conscious of their debt to Owen. Frost runs over a set of innovations—public baths and washhouses, improved working-class housing, and reformatory and industrial schools—which can plausibly be traced to Owenite policies and practice:

a little consideration must satisfy every impartial mind . . . that the institutions mentioned have sprung directly from the conviction that man is the creature of circumstances.

[41] B. Taylor 1983: 62, 301; J. S. Mill 1989: 134.

So numerous have been the reforms of this kind during the last twenty or thirty years, that they can be ascribed only to the gradual awakening of society to the conviction that there was much more in Socialism than its critics were willing to acknowledge.[42]

As I have stressed, there is no logical connection between the belief that people are moulded by their circumstances and an anti-sensual sexual ethic. One may believe the former and be a thorough-going sexual hedonist and libertarian, without inconsistency. For environmentalism to bolster anti-sensualism it is necessary that the latter position is found persuasive independently (but if it *is* found persuasive, environmentalism—above all with its suggestion that the libido is subject to external modification—becomes a powerful rein-forcement). Owenism is especially interesting because in several respects it offered a clear invitation for the wedge to be driven between environmentalism and anti-sensualism.

Owenite determinism, as worked out in detail, was an arbitrary and incoherent doctrine, picking and choosing what counted as 'cir-cumstances' without much rationale. In the sexual domain one of the inescapable formative conditions of an individual's character was taken, rather oddly, to be his or her sexual preferences. These can change in a way that is outside our control or accountability, said the Owenites. All agreed that, in a negative form, this argument justified the divorce of couples who had fallen out of love; but positively, it clearly might allow for the free formation of a succession of unions.

Another respect in which Owenite socialism threatened the nine-teenth-century linkage of environmentalism and anti-sensualism was its thinking on the family. Owen's co-operative vision, developed at New Lanark in the second decade of the century, had at first attacked competitiveness and self-interest in their economic and class guises, and was then broadened in the 1820s to include social arrangements, especially the closed unit of the indissolubly married couple and their children. Owen spent the years 1824–9 in America, where his main effort was to set up a colony of co-operative communities at New Harmony, Indiana. When he returned to England he directed his energies for some years to forging a grand universal trade union, but when this venture collapsed Owen turned back to the project of a new social architecture. As a result several Owenite communities were launched in the British Isles.

[42] Tholfsen 1976: 116; Frost 1880: 22–3.

To make the community the overriding unit was to place men and women in a force field which might be stronger than any bond they formed with a single partner. The communitarian ideal was thus ambiguous in import; the renunciation of family might amount to the adopting of a celibate regime, and I shall mention later the links between the Owenite experiments and such anti-sensual groups as the Shakers. The anti-sensual strains in socialism (which certainly existed) were significantly different from the anti-sensualism of William Godwin or James Mill, in the sense that socialists do not seem to have envisaged a reduction in the strength of human sexual drives. This was consistent with their idiosyncratic way of drawing the line between environment and character, such that an individual's sexual preferences oddly belong to the first of the two domains. In so far as these preferences are understood to be a preference *for* another partner (and not just a preferring not to cohabit with an existing one), there is an assumption that the drive to sexual union is constantly present, and of course unmodifiable because not part of personal character.

William Thompson, an important though independent-minded Owenite theorist, I have noted earlier in this study as a firm believer in the evils of female sexual abstinence. In his plans for a socialist community Thompson envisaged unrestricted casual intercourse, with no marriage in the orthodox sense. (He was unusual among progressive thinkers in accepting the Malthusian arithmetic of population growth—and thus anticipating George Drysdale a generation later—but his eccentric confidence that a woman could not conceive unless she remained prone in bed for some time after intercourse allowed a neat reconciliation of the creed of casual sex and the requirement of fertility control.) Thompson's thinking about communitarian sexual arrangements seems to have been pondered sympathetically by Owen—perhaps, to judge from a rather puzzling anecdote (dating to 1832) by William Lovett, with too much eagerness even for Thompson's taste.[43] Owen was apparently backed by

[43] W. Thompson 1830: 231–46; G. J. Holyoake 1875: ii. 412. Here is the anecdote, at Lovett 1876: 49–50: 'the proposal to establish an incipient community upon Mr Thompson's plan was opposed and marred by Mr Owen. It was in this curious manner. After the proposal was discussed for some time . . . we retired for dinner. When we came back our friend Owen told us very solemnly, in the course of a long speech, that if we were resolved to go into community upon Mr Thompson's plan, we must make up our minds *to dissolve our present marriage connections, and go into it as single men and women.* This was like the bursting of a bomb-shell in the midst of us. One after another, who had been ardently anxious for this

George Petrie in a dispute about the need for communitarians to renounce their existing spouses. Petrie was one of the heirs to Thomas Spence and his programme of agricultural co-operatives, and is known to have envisaged completely unregulated liaisons in Spencean communities, with a woman's sexual needs especially in mind.

He went so far as to denounce 'moral marriages'—that is, unofficial but solemnly undertaken liaisons by non-communitarians—as a worse expedient than conventional marriage as long as the old social structures were in place. Petrie's reasoning seems to be that such arrangements would still stifle a woman's sexuality, while actually putting her in a more dependent position than a legal wife. Petrie may have especially in mind the 'moral marriage' which Richard Carlile and Eliza Sharples had proudly announced to the world at this time. 'A marriage more pure and moral was never formed and continued in England', they declared. The Carlile-Sharples union was admittedly a marriage of 'two bodies' as well as of 'two congenial spirits', so here the vocabulary of purity and virtue should probably be linked to the Owenite rhetoric of non-continent chastity, which I discuss shortly. But Petrie would certainly have been disgusted by the entirely non-physical moral marriage which John Stuart Mill and Harriet Taylor had undertaken at almost the same date.[44]

Robert Owen apparently agreed with the anti-chastity views of his son, Robert Dale Owen, at this period (though he was notoriously silent concerning influences on his thought). It may have been a case of son teaching father, as the two men co-operated closely in several important projects such as the New Harmony colony and the editorship of the *Crisis*. Robert Dale was 25 when he joined his father in America. He became responsible for New Harmony's magazine, the *Gazette*, and in his journalism of the next few years it is evident that his views on sexual issues were taking an increasingly confident and progressive form. The radical Frances Wright, who may or may not

proposal of a community, began to express doubts, or to flatly declare that they would never consent to it; while others declared that the living in a community need not interfere in any way with the marriage question. One poor fellow, Mr Petrie, an enthusiast in his way, quite agreed with his brother Owen, and made a speech which many blushed to hear, and contended that it would make no difference, as he and his wife were concerned, for she would follow him anywhere . . . nothing could have been better devised than this speech of Mr Owen to sow the seeds of doubt, and to cause the scheme to be abortive; and when we retired Mr Thompson expressed himself very strongly against his conduct'.

[44] Saville 1978; *Man*, 1 (1833), 97–8; *Gauntlet* (1834), 521–2; *Isis* (1832), p. v.

have been his mistress but was his companion on American and con-
tinental journeys, was an important coadjutor in this process.[45]

In 1831 Robert Dale published *Moral Physiology*, a very early
defence of birth control which was to acquire an almost legendary
status as a result of the campaign to issue such literature fought in the
1870s. Robert Dale had favoured the cause of birth control for some
years, but had hitherto refrained from going into print on the matter.
During his co-editorship of the *Crisis* with his father, in the course of
a return visit to England, the magazine carried advertisements for
Moral Physiology, though Robert Owen's attitude to birth control is
another obscure feature of his sexual philosophy at this period.
Francis Place claimed to have been shown sponge contraceptives by
Owen in 1818, who had acquired them in France with a view to deal-
ing with any population problems at New Lanark by this means (a
policy also mentioned by Holyoake, many years later). Owen later
denied a version of this story, but it is hardly possible that Place was
wrong in his recollection of getting some information about barrier
methods from Owen. Most probably Owen, who did not accept
Malthus's arguments, was keen not to appear to be a friend of population control. Although his positive advocacy of birth control may
have ceased by 1820 there is no hint in the next twenty years that he
had become hostile to such practices.[46]

The practical advice Robert Dale offers in the *Moral Physiology*
(which is virtually confined to coitus interruptus) is not as significant
as the general attitude to sex conveyed in this attractive essay. Robert
Dale's position is at once perfectly hedonistic and in favour of temperance, an apparently difficult combination which he makes remarkably convincing. He insists that sexual satisfaction for its own sake, in
both sexes, is not only necessary but positively valuable (it is admitted as a possible objection to coitus interruptus that male sperm in
the vagina may be necessary for a woman's orgasm). But beside this
core of pro-sensual principle are set countervailing views about
which Robert Dale is no less emphatic. Sexual dissipation is an even
worse evil than celibacy. To sexual vulgarity, such as dirty jokes, the
author is happy to confess to being 'girlishly sensitive', Orthodox

---

[45] Leopold 1940: 37–68, *passim*; R. D. Owen 1874: 266–7; *Co-operative News*, 35 (1904),
465–6, 494.

[46] Himes 1930*a*, 1928; B. Taylor 1983: 299; Miles 1988: 140–3; Peel 1963; Boner 1955:
44; Field 1931: 98–102. See also the letters from William Hawkes (31 Mar. 1831) and
Francis Place (4 Aug. 1832) in Owen Collection, Holyoake House, Manchester.

marriage is often *too* sensual, once priestcraft has given its permission for intercourse: it might be a good thing if spouses sometimes slept in single beds.

The *Moral Physiology* thus discloses interestingly a logic or pattern of attitudes which suggests how it was that so many Owenites maintained, at least until the late 1830s, their unusual reconciliation of pro- and anti-sensual beliefs. For example, Robert Dale has a good deal to say both for and against Shakerism, and the Shakers are used by him at times virtually as a means of self-definition. Shaker celibacy is deplored, of course, but for interesting reasons. It reveals an excessive interest in sex, according to Robert Dale, and is actually just the converse of licentiousness: in general, negatively or positively, we are too preoccupied about the subject. Here Robert Dale in fact concedes some degree of cultural influence on our sexual drives, in a partial step towards Rousseau's position.

Above all the Shakers are wrong, in Robert Dale's view, to regard the sexual instinct as 'a "carnal and sensual passion"': 'it appears to me the most social and least selfish of all our instincts'. The contrast, or non-contrast (carnal/sensual versus social/unselfish), is very telling, and very Owenite. The most distinctive and, it must be admitted, strange passages in the *Moral Physiology* are those celebrating sex as an altruistic and benevolent social agency, as 'a means by which the sexes mutually polish and improve each other'. But such thinking was in tune with that of other socialists: William Thompson called the sexual instinct 'this most innocent and kindly-disposing of human passions'. As late as the mid-1840s, when libertarian Owenites were much embattled, a man of this persuasion urged in the Owenite press that 'philanthropy ... finds its highest expression, and is ... brought to a focus in the free and unthwarted union of individuals of different sexes'.[47]

I have been seeking to show that there was much in the conceptual scheme of Owenism to encourage a splitting of environmentalism from anti-sensualism, and even a novel conjunction of an environmentalist mode of thought with pro-sensual codes. Robert Owen seems to be drawn to such possibilities in his social thinking around 1830. That they advanced little further towards realization is a considerable testimony to the power and pervasiveness of the anti-sensual mentality in nineteenth-century England. For the Owenite sexual revolution, to judge by the record of events in the 1830s, was

[47] R. D. Owen 1877: 11–13, 15, 33, 42, 53, 63–4; W. Thompson 1830: 246; *New Moral World*, 13 (1845), 375.

defeated both from within and from without, and by both theoretical and practical pressures.

This story, until quite recently a forgotten episode in English radicalism,[48] is not always an easy one to piece together. The main surviving record of socialist policies and attitudes, the Owenite printed literature, is only part of the picture, for much was transacted in lectures and discussions. This literature comes in a variety of forms and from various stables, and it is to some extent a matter of conjecture who would have encountered it, and how representative of socialist beliefs it was. For there were certainly many points of view within Owenism in the 1830s about the right approach to sexuality, with a somewhat artificial impression of unanimity being given by Owen's personal ascendancy in most, but not all, of the published Owenite propaganda. There is also some descriptive record of such things as socialist lectures, mainly from the enemies of the movement, who could be astonishingly rancorous. Even the non-*parti-pris* local press tends to be extremely hostile to socialism.

From the outset, a weakening of the existing institution of marriage was part of Owen's programme at New Harmony. Local scandal, and probably Owen's reputation in America generally, had it that sexual promiscuity (with the leader participating) became the order of the day in the colony. In fact it is likely that Owen favoured dissoluble monogamous unions among the residents at New Harmony, involving a declaration of loyalty by the partners, and any further procedures necessary to avoid illegality under local laws. Intercourse was apparently permitted between couples who had announced an intention to marry in this sense, and there was an attempt after marriage to prevent families becoming too autonomous in the community. Back in England, in 1830, Owen was possibly attracted by William Thompson's bold vision of communitarian sexual arrangements, as I have mentioned, and he discussed marriage in lectures of that year with a great emphasis on the evils of the existing state of affairs. There are only hasty and 'intermediate' proposals for a reformed system, but this in itself adds to the libertarian feeling of the statement.[49]

---

[48] For its progressive disclosure see J. F. C. Harrison 1969: 59–62 (and bibliography of Owenite and anti-Owenite literature pp. 266–77, 286–358), Saville 1978, and B. Taylor 1983: esp. chs. 2, 5, 6, 8.

[49] Muncy 1973: 54–9; Bestor 1950: 222–3; Willard 1833: 308; B. Taylor 1983: 337; R. Owen 1830: 83–4.

In May 1833 Owen promulgated his alternative mode of marriage more emphatically. In the ultra-tidy new scheme a couple were married when they confirmed a declaration of intent to cohabit monogamously which they had made three months earlier. They might divorce, after not less than a year of marriage, through a similar process of declaration and redeclaration, only in this case six months must separate the two ceremonies.[50] This is certainly interpretable as a formula for committed monogamy, however short-lived and unofficial, and it was cited continuously against accusations that Owen's was a free-love system.

But its full moral implications hinged very much on the role of community. If this form of 'marriage' is envisaged as applying only to unions formed by the hitherto unmarried in the utopian social climate of a socialist community it is indeed unalarming. It does not seek to interfere with conventional marriage outside the community, or require that already-married couples entering the community change the footing of their union. Its worst feature is the prospect of new liaisons every two years, but, as Owenites often argued, young unmarried men and women brought up in a community would surely be able to make well-founded declarations of love, as durable as the despotic 'circumstances' of our sexual tastes permit.

In reality the scope of the Owenite sexual programme was not as modest as this, or at least not with all Owenites at all times. In the mid-1830s Owen himself was so seized by confidence in the power of the community idea to transform human society that the distinction between communitarian and non-communitarian forms of life became of little importance to him. Within months, he kept asserting, the whole of England would be a community. The journal he founded in 1834, the *New Moral World*, bore a title expressing exactly this confidence. It was Owenite women, prophetically, who were troubled by the consequences of an insouciance about the sexual codes of the old moral world. One wrote to Owen's earlier magazine, the *Crisis*, claiming to speak for many other women and uttering a passionate warning against jumping the gun on socialist sexual unions: until new codes were really in place women would suffer enormous penalties if they renounced conventional marriage. Owen's first reaction, appealing to the imminence of the new order, was dismissive, but within a few months the *Crisis* was insisting that

[50] *Crisis*, 2 (1833), 143–6.

it did not give its blessing to unofficial marriages between Owenites 'in the present circumstances of society', and recommending that a solitary partner in an existing marriage should never be admitted into a community (it was another anxiety of Owenite women that their husbands might desert them for community life).[51]

But as the *Crisis* reveals, and Thomas Frost recollected, some Owenite men and women *were* deciding to 'anticipate the natural freedom of the two sexes'. Details are recorded of four such 'marriages'—mutual public pledges of fidelity—between Birmingham couples in 1834. As late as 1842 two men and a woman in Cheltenham grafted a similar ceremony on to a private separation agreement (a traditional mode of informal divorce which still thrived at this period): thus dissolving one (orthodox) marriage and creating a new (unorthodox) one, with the assent of all parties.[52] For all its new-found caution about informal marriages the *Crisis* in 1833 could not bring itself to be wholly sceptical of the couples' chances of happiness: 'they may be happy—happier than if the priest had yoked them'.

Here is voiced one of the most distinctive and no doubt potent themes in socialist sexual reformism: a kind of horror at the alleged quality of relations between the sexes under the existing rules. When Owen eventually drafted a full statement of his new ideas it was in a series of lectures delivered at Leeds in 1835, subsequently to become notorious, the *Lectures on the Marriages of the Priesthood of the Old Immoral World*. The rather preposterous title accurately conveys the vehement negativity about orthodox sexual relations from which Owen's policies sprang, better than the title generally adopted for later reprintings of the series, *The Marriage System of the New Moral World*. Owen's antagonism to orthodox heterosexual unions, which runs at least from the 1830 lectures and probably from New Harmony days, produces by far the most vigorous and unusual passages in all his surviving comment on the marriage question. In its climactic expression, in these 1835 lectures, Owen speaks as a man charged with a duty to transmit a vision which has been granted to him, the vision of a great key to his society's troubles: its sexual arrangements. He is at once Old Testament prophet and harbinger of the twentieth-century stereotype of his era as essentially disordered in its sexuality. This element is supposed to be on the same footing as

[51] Armytage 1961: 141; Podmore 1906: 462; *Crisis*, 2 (1833), 190, 192, 3 (1833), 100, 143.
[52] Frost 1880: 16; *Crisis*, 4 (1834), 54, 108; B. Taylor 1983: 198–9; Stone 1990: ch. 7.

certain other social, administrative, and economic aspects of the society, but in the 1835 lectures sexual issues are given a primacy:

Not to speak and expose the greatest of the practical sources of vice and misery! As well may we at once close our eyes and put a seal upon our lips, that we may not see or speak of the greatest deceptions, the most abandoned wickedness, and the cause of the most extended calamities known to the human race. The time is now past for those who have a desire to ameliorate the condition of the human race, upon everlasting principles of truth, to pander to the prejudices of any portion of mankind; much less to this prejudice in favour of artificial marriage and unnatural union of the sexes: of a prejudice which is one of the deepest seated in human society, and the most fatal to the happiness of men and women from their birth to their death.[53]

Though no other socialist, perhaps, used quite Owen's vocabulary about matrimony as a 'satanic' and 'accursed' institution, or offered such harsh images of discord and frustration as the condition of almost all modern marriages, protest about the present state of affairs runs right through the literature: about the life of the married and, conversely, about the scale of prostitution, which all socialists agreed was the outcome of marital unhappiness. Lecturers such as Joshua Hobson seem to have found it extraordinarily hard to move from attack on the old to the account of the new which they kept promising, and *The Times* complained that in a lecture by Charles Southwell on marriage 'a more numerous collection of "negatives", in all probability, was never given utterance to within a similar space of time'.[54]

The rhetoric was negative, but no less emotionally telling for that, and capable of stirring Owen's followers to 'take the liberty of applying for themselves' his doctrines in a radical way. Not only the enemies of socialism believed that Owen 'had recommended . . . promiscuous intercourse and laxity of marriage tie', but also the occasional well-wisher, who had to be put right by official Owenism. The segregation and independence created by the mechanism of the community—independence even from Owen's powerful paternalistic sway—offered the opportunity for a radical application of Owenite sexual theory. Owen had said in the 1835 *Lectures* that a socialist community would not work if it had to 'consist of married couples', but had not been more specific. At Manea Fen, in Cambridgeshire, a community was founded in 1838 which did operate free-love poli-

---

[53] R. Owen 1839: 8–9.   [54] Hobson 1838: *passim; The Times* (23 Aug. 1839), 5.

cies, the land being made over by the founder, William Hodson, on the express condition of 'personal freedom' for the colonists.[55]

Manea Fen quickly collapsed, like all the Owenite communities. It had always been discountenanced by the official Owenite administration as a maverick venture, but even its well-wishers, who had started a periodical called the *Social Pioneer* to report achievements at Manea Fen, had to condemn the community as a 'worthless and polluted thing' when its libertarian policies became apparent: 'our great and glorious cause must not, shall not, be allowed to sink, through the follies and indiscretions of a few unreflecting individuals'.[56] The blend of ideological and prudential considerations is very characteristic of socialism as it sought, in the late 1830s, to repudiate the libertarian potential of the views on sexuality which its leader had brewed for ten years and then uttered so arrestingly in 1835. The retreat was wholesale and rapid, but it is a fine question how spontaneous it was, and how much enforced—and, if enforced, by which agents.

William Hodson and his associates claimed that they were applying Owen's own teachings at Manea Fen, and were confident enough about this to apply to him personally for his endorsement. But they were wrong in so far as Owen in his public pronouncements always envisaged a high level of sexual exclusivity, and in that sense restraint, as flowing from the abolition of the old priestly marriages. Coupled with his continuing belief that abstinence was an evil, and 'CELIBACY . . . a great crime, necessarily leading to diseases of body and mind', the result is an idiosyncratic rhetoric whereby 'chastity' is redefined as affectionate intercourse: 'the pure and genuine chastity of nature is to have connection only with affection'. Chastity becomes, in fact, a constituent of the sexual experience of a couple: 'chastity . . . exists only between the sexes when in their intercourse they feel a sincere and genuine affection for each other'.[57] Among Owen's followers the formula, chastity = affectionate intercourse, became habitual, and their leader's notable emphasis on purity, refinement, and high-mindedness in sex, however idiosyncratically the notions were understood, was valuable in rebutting accusations that Owenites wanted rampant promiscuity.

[55] R. Owen 1839: 7, 21–5, 71; *New Moral World*, 6 (1839), 984, 10 (1842), 151; Armytage 1961: phase ii, ch. 8; B. Taylor 1983: 253–7; *Social Pioneer* (1839), 63.

[56] *Social Pioneer* (1839), 44–5.

[57] B. Taylor 1983: 256, 347; letter from S. Rowbotham (20 Apr. 1839) in Owen Collection, Holyoake House, Manchester; R. Owen and Legg 1839: 12; R. Owen 1839: 10–12, 49.

In a strange further extension of the idea of the 'chastity of nature' most disciples did not feel able to follow Owen. He speaks in the 1835 *Lectures* of animals as more 'chaste' than Man, the latter being, confusingly, 'a mere irrational animal' in so far as he doubts his own capacities for affectionate sexual fidelity. Logic was never Owen's strong point, and his opponents seized on the favourable remarks about animals, while giving the 'animal' a conventional pejorative sense (as Owen himself does on his second use: the paradoxes of his rhetoric were pointed out in several attacks). Owen's was labelled a 'farmyard morality', a matter of 'bull and cow', and there is sometimes a sense of quite genuine repugnance at the introduction of the comparison with animals (one anti-socialist felt as if he were smelling a skunk).[58] Whatever Owen may have had in mind about the moral refinement of animals (and the *New Moral World* did carry occasional stories about their mental powers) his followers generally disavowed any 'animal' element in socialist sexual policy. The 'natural', for human beings, meant the 'rational', and Owenite missionaries such as Lloyd Jones could talk themselves enthusiastically into the same moral universe as their opponents:

The nature Mr Owen is speaking of, is human nature, and that in its best form—educated human nature . . . so soon as the finger of cultivation touches your mind . . . no desire will arise but of a pure and elevated character . . . as you ascend the scale of reason, mankind take up a different position, and they will never approach the brute.

The antagonist in this particular debate, the Reverend John Bowes (who wrote an anti-Owenite pamphlet entitled *The Social Beasts*), understandably observed, 'Mr Jones is on my side, and we are both against Mr Owen'.[59]

Not all the converging movement in these cases came from the Owenite forces, even if some of the avowedly religious opponents of socialism did have a lively sense of depraved and fallen human nature (like the biased chairman of one debate who recalled that 'early in life, I had to contend with strong passions, which, had I been a socialist, must . . . have inevitably hurried me into an abyss of

[58] R. Owen 1839: 40–1; Brindley 1840: 4; Bowes 1840: 4; Giles 1839b: 58–60, 72; Campbell and Roberts 1838: 39; R. Matthews 1840: 3, 14.

[59] Buchanan 1841: 20; *New Moral World*, 9 (1841), 230; R. Owen and Brindley 1841a: 43; Bowes and Jones 1840a: 67, 88, 92.

sin').[60] In others antagonism to the perceived licentiousness of socialist doctrine sprang from an aspirational anti-sensualism not very far removed from the secular creed of their Owenite opponents (and in the case of this confrontation, as I have mentioned, there was no belief in the decay of human libido, on the progressive side, to sunder the parties). One anti-Owenite was proud, indeed, to steal terminological thunder and call himself a 'Christian Socialist', an early use of the expression: 'the improvement of mankind, intellectually and morally, is the great purpose of the gospel'. Another clerical attacker makes it sound as if it is the Owenites who are Calvinists by virtue of their views on modern marriage (that 'state of bliss'): 'could we, for a moment, suppose human nature to be so corrupt?' In the calculated appeals made to Owenite women by such writers and speakers, which I shall discuss shortly, the role of Christianity as an alleged raiser of women's status was often mentioned.[61]

It is a nice irony that in 1836, with the introduction of marriage by a public registrar, English marriages in the old immoral world ceased to be exclusively priestly. This offered an important opportunity of reconciliation between Owen's hopes for marriage, and the contemporary reality. By the end of the decade he was consistently claiming that his only surviving quarrel with English marriage law was the lack of provision for divorce. Already in 1837, in lectures, he had mentioned socialist marriage policy with a much greater emphasis than hitherto on its capacity to yield stable and happy unions. His platform utterances now, it was said, gave little hint of the views he had espoused two or three years earlier.[62] In 1839 Owen started openly to avow revised doctrines on sexual relations, and for the first time the signs of a socialist climb-down on sexual policy were acclaimed by his opponents.

The earliest written and spoken attacks by outsiders on this aspect of the Owenite system date from the previous year, but it is in April 1839 that rumours of bold goings-on at Manea Fen start to emerge, and these may have been the essential catalyst for the official retreat.[63] In May Owen gave an address to the socialist Congress

[60] R. Owen and Brindley 1841b: 98; also Whalley 1840: 4–13 (the author was a converted atheist); Mather 1839: passim; Dalton and Jones 1838: 28; Heath 1838: 21.

[61] T. H. Hudson 1839: 238; Sturmer 1840: 11; Scoresby 1840: 9; Campbell and Bannister 1839: 87.

[62] R. Owen 1837: 77-8; Buchanan 1841: 38; Giles 1839a: p. iii.

[63] Beard 1839: 150; Birch 1839: 6; New Moral World, 6 (1839), 963; Social Pioneer (1839), 44.

which contained remarks on marriage hailed by the *New Moral World* as timely when 'we are assailed alike by open opponents and pretended friends, upon the latter subject . . . and the stability and progress of our Association threatened'. In the address Owen too had spoken anxiously of the difficulty of conveying the socialist view on sex to the unconverted (the 'most difficult task which your missionaries have to perform'), and of the dangers of subversion from within: 'on no account abandon regulated marriages and divorce, or listen to those who recommend promiscuous intercourse in your societies. My present impressions are, that for ever there must be rationally devised marriage and divorce'.

These are eloquent remarks. There is surely some regret that a more thorough revision of sexual codes is not in prospect, but the concessions to practical realities are firmly insisted on. 'The subject must be fairly met in principle', says Owen, 'but the practice for the present irrationally trained generation must be adapted to the present state of human intellects and feelings'. He is also tellingly inconsistent as to the duration of the period of compromise on sexual 'principle'. Is it a 'present' necessity, or will it last 'for ever' (as his 'present impressions' tell him)? And one could fairly observe that the 'principle' of Owenite sexual thinking was itself equivocal, with its strange notion of an incontinent chastity, and suggest that a liberationist experiment at Manea Fen had forced a commitment to one strand of this creed. Owen never squared the relationship of creed and chronology, perhaps because the former was so ambiguous. Twelve years later he was still fiddling with the question, at once downgrading the Owenite ceremonies of marriage and divorce as 'regulations to be observed . . . in the early stages of the transition from the present deplorable conditions of society to the coming consummation of human goodness and happiness', and at the same time declaring that 'it is in no way proposed to apply these views or regulations to the present system of society'.[64]

In 1840 the attack on Owenite sexual policy moved to a new level of intensity, in the wake of a much-noticed series of exchanges in the House of Lords early in the year. The prime mover here was the principled reactionary Henry Philpotts, Bishop of Exeter, who was urging the government to institute some kind of inquiry into the workings of the socialist movement. In the next two years clashes

---

[64] *New Moral World*, 5 (1839), 440–4; *Robert Owen's Journal*, I (1851), 196.

between socialists and their opponents became much uglier, with actual violence threatened against Owenites, and committed on property.[65] Episodes in this 'Holy War', as Owen's biographer has christened it, were often associated with the sexual issue: triggered by a lecture or debate on the subject, or centred on individuals supposed to hold liberationist views. Owen himself became notably defensive on socialist sexual policy, publicly repudiating, for example, the views of his own son in the *Moral Physiology*, and reissuing several previous statements on marriage and divorce so packaged as to demonstrate that he had always been a friend to marriage, in 'hearty concurrence' with the enemies of 'immoral principles and practices'.[66]

While Owen was striving to present an undismaying image of his views, other campaigners for socialism, particularly the 'social missionaries' (or paid-up local proselytizers) appointed in 1837 and 1838, were seeking to allay alarm about the sexual policies of their organization too. Frederick Hollick, the Glasgow missionary, seems to have tried to keep the subject out of public debates and lectures as much as possible. Other missionaries, when they did declare themselves, went to great lengths to conciliate hostility on sex and marriage, completely repudiating Robert Dale Owen ('he does not belong to us'), making out that marriage of informally united couples would actually be *required* for entry into a community, and even purporting to look forward to the elimination of divorce (one of the hitherto secure planks in the socialist platform) once the new order had been in place for a time.

One missionary argued that, far from being out of step with public opinion on sexual questions, this was the area (despite all the 'odium') where socialist belief most 'nearly coincides with that of every good and intelligent mind'. There was generally a tendency to vanish, chameleon-like, into sexual orthodoxy: a supporter wrote quite bluntly that the whole purpose of Owen's policy was 'to put an end to promiscuous intercourse, and to strengthen and extend the institution of marriage, and of domestic purity, and female chastity'. Some confidence on the part of Owenites that they were in tune with ordinary opinion is shown in the fact that they occasionally resorted to

---

[65] L. Jones 1895: 360–72; Podmore 1906: 511–22; B. Taylor 1983: 185–92; Royle 1974: 64.
[66] Owen and Brindley 1841a: 23; see Owen's *Lectures on the Marriages of the Priesthood of the Old Immoral World* (1840) and *Manifesto* (1840); *New Moral World*, 7 (1840), 1156.

satirical pastiche of their enemies as a means of counter-attack. There was also a certain amount of prim complaining about the 'obscenity' of the language used by anti-Owenites, and about the 'offensive sounds' from hostile mobs, 'which, happily, I could not understand'.[67]

Given Owen's history of bold policy-statements a good deal of this defensive work took the form of favourable exegesis of, or counterquote from, or straight denial of the authenticity of, writings associated with Owen. In the last case there sometimes came to light interestingly an apparent element of pluralism within socialism, though it was still to an extraordinary degree one man's movement. A particular cause of irritation were the first three editions of the 1835 *Lectures* published by Joshua Hobson in Leeds, the text of which was repudiated as being unauthorized 'abstracts' made by a reporter, and which contained as an appendix a quotation from Shelley's notes to *Queen Mab*, put in to make up space. The Shelley note, which I have cited in an earlier chapter, is on the face of it close in spirit to the bolder Owenite views on sex, which it was surely an inspiration for. Owen and others publicly deplored the inclusion of this 'offensive and obnoxious' passage.

If Hobson, who was the publisher of the *New Moral World* and a trusted and effective Owenite, and who himself lectured on socialist sexual policy in a loyally defensive if vacuous way, was responsible for the Shelley appendix the hostile reaction from colleagues would appear to be a matter of rattled nerves. After all, the same text of the lectures had appeared in the *New Moral World* (and even the authority of this organ of the socialist establishment was sometimes denied by defensive missionaries). But Owen said the appendix had been inserted by 'one of his Committees in London'. Holyoake regarded it as provocative in itself to release the lectures from their obscurity in the pages of the *New Moral World*, and judged that some individual 'on mischief bent' must have contrived for Hobson to make them more widely available. Certainly somebody in Manchester in 1835 made trouble for socialism by issuing an edition of Owen's *Outline of a Rational System of Society*, of 1830, entitled *The Social Bible* and enlarged with declarations about how 'the irrational sounds of husband and wife . . . will not be heard' in the socialist utopia. This

[67] Brindley and Hollick 1840: 9; Scoresby 1840: 7; Troup and Jones 1839: 4–5; Buchanan 1841: 45; *New Moral World*, 9 (1841), 75, 230, 402; T. Hall 1841: 6; 'A Clerical Gentleman' 1840: *passim*; 'Publicola' 1844: *passim*.

production also led to embarrassment and excited repudiation among Owenite campaigners in the difficult years around 1840.[68]

The most significant fact in this confusing scene is probably that the first recorded attack on *The Social Bible* came much earlier, at the time of publication, and from *within* Owenism. William Hawkes Smith, a prominent Birmingham socialist and Unitarian, thought his Manchester colleagues had adopted a 'bad policy' in using 'terms as discordant as possible from the prepossessions and present feelings of mankind'; 'the wisest course' was 'to point out the various, the *numerous* instances in which all that is most sublime and beautiful in *existing* systems, is, in fact, selected and brought into harmonious action in the *New* System'. Though this purports to be strategic advice from a man who 'would never *blink*' an unpopular opinion it is clear that the author genuinely perceives an identity between the goals of Owenite sexual policy and those of orthodox morality at its finest. What libertarian element he does concede in the former is nervously utopian: 'it may be . . . that in the *perfected* state of society contemplated . . . attachments would be so pure—would be in every case so *certainly* the sole motive for the connexion, that *ceremonies* might become needless'.[69]

Hawkes Smith went on to lament in 1839 how Owen had 'injudiciously' left socialist marriage doctrine 'open to a double and unfavourable construction', and to advise Owenite campaigners to steer clear of these 'thorn-bushes'.[70] But he had given very similar counsel in 1835, some three years before anyone outside the movement had thought to attack this doctrine, which must tend to call in question the importance of the attack, when finally it came, as a cause of the Owenite climb-down. I have already illustrated the inherent ambiguities of Owen's own attitude, and I shall identify shortly several active if fringe Owenites, and one social missionary, who are known to have been strongly attracted to anti-sensual moral systems. How convincing is it that the 'Holy War' would have forced a loyal Owenite to withdraw from a libertarian posture on sexual questions?

The hostility which the Owenites encountered in the years 1838–42 as they travelled the country lecturing and debating was

---

[68] Buchanan 1841: 23; Owen and Brindley 1841a: p. xvi; Owen and Legg 1839: 12; *New Moral World*, 5 (1839), 519, 6 (1840), 723; G. J. Holyoake 1875: i. 211; Troup and Jones 1839: 4–5; R. Owen 1835: p. xlii; Campbell and Bannister 1839: 51.

[69] On Smith see J. F. C. Harrison 1969: 74; *New Moral World*, 2 (1836), 333–4.

[70] W. H. Smith 1839: 45–6.

certainly often ugly and dismaying. One of the climactic occasions was Owen's three nights of debate with the schoolmaster-cum-self-appointed scourge of socialism, John Brindley, in Bristol in January 1841. Even the Owenite account of this admits the degree of hostility in the audience, and Owen appears to have been seriously discon-certed and put off his stride by the barracking (he was, after all, a man of 70 by this date). There was a further week of disruptive and threatening action against other socialists in Bristol later in the month. But the *New Moral World* claimed that the hostile groups and crowds in these events, however large and however favoured with a blind eye by the authorities, were just elements stirred up by Brindley; in reports of other clashes it is apparent that the heckling and hissing supporters of Brindley or Joseph Barker were a well-defined group in an audience, perhaps sitting round their man.[71]

Owenite sexual policies were brandished inflammatorily on all these occasions, and it would be true to say that the most aggressive anti-Owenites were also the most obsessed about the socialist theory and practice of sex. Less virulent enemies also clearly felt that this was a strategic point at which to attack the convictions of working-class socialists and would-be socialists, a potential fault-line between elements in the Owenite creed. 'The question of religion is not prominent here', said a speaker at a lecture on marriage to which 500 socialists had been specially invited, 'it is the existence of morals, the sacredness of our persons, the peace of our relationships'. But were Owenite campaigners right to feel that 'calumny and misrepresenta-tion' on sexual policies 'has threatened to sweep us away in the stream of public opinion', and that popular sensitivity about sex would cause socialism to be perceived as a 'danger' by those whose loyalty was to be won?[72]

To start with, as a modern historian of this episode has com-mented, lectures and debates on marriage and sex were 'crowd-catchers', and must have spread socialism's message even if they also increased the number of its enemies. (The programme of a debate-series was on occasion spontaneously altered to accommodate the audience's demand for more discussion of controversial sexual ques-tions.)[73] Secondly, sex did not figure as much in the campaign against

---

[71] Owen and Brindley 1841a: *passim*; G. J. Holyoake 1875: ii. 239–41; L. Jones 1895: 360–6, 368–72; Podmore 1906: 520–2; *New Moral World*, 5 (1839), 196, 7 (1840), 1284, 9 (1841), 37, 66; *Antidote to Mormonism and Infidel Error* (8 Aug. 1857), 55.

[72] Robert Matthews 1840: 57; Eyre 1840: 14; Hobson 1838: 79.

[73] B. Taylor 1983: 140, 185; Bowes and Jones 1840a: 69.

Owenism, even in the period 1838–42, as some Owenites seem to have perceived. Rather betrayingly, a writer in the *New Moral World* in 1843, looking back on early 1840 and the stir caused by Philpott's House of Lords attack, recalled that 'the foundation of all this out-cry—that which constituted the essence of the charge against our views, was that portion of them which had reference to Marriage and Divorce'. This, like Thomas Frost's much later recollection of these events, is quite misleading as to the balance of Philpotts's argument, which was comprehensively opposed to all aspects of socialism as the creed of a dangerously large, vigorous, and well-organized subversive political movement.[74]

It is possible that 'socialism', for a middle-class individual around 1840, strongly connoted 'a new and revolting system of morals', or common-law marriages, temporary concubinage, and exchange of partners among the working class.[75] But the more knowledgeable attacks do not necessarily give much attention to the sexual element in the Owenite programme (just as Flora Tristan, reporting on social-ism to her French readership in 1840, does not touch on this, while a less well-informed compatriot, a year later, acquires the impression that completely free unions of the sexes are the main socialist doc-trine).[76] There was some feeling even among the authorities that John Brindley was too aggressive in his methods. The country's press, from local papers up to national quarterlies, was uniformly hostile to socialism, but quite often showed a concern not to be a party to unbalanced views of Owenite sexual policy.[77]

Owenites had an admirable appetite for opening up their doctrines to popular discussion, and perhaps a canny sense that all publicity was good publicity. The original masthead on the *New Moral World* said this about socialism: 'Silence will not retard its progress and opposition will give increased celerity to its movements'. It was the Owenites who invented the device of the series of nightly debates in provincial centres, which had an extremely important continuation with Bradlaugh's free-thought movement. In the period around 1840 their nerve seems to have failed, even though Philpotts's activity in the Lords was recognized by outsiders to have told 'the world' about

---

[74] *New Moral World*, 7 (1840), 1209, 11 (1843), 369; Frost 1880: 17, 20.

[75] Brindley and Hollick 1840: 9; PP 1840, xxiv. 90; *Sanitary Review and Journal of Public Health*, 3 (1857), 328. See also the satire on 'randy' socialist girls in Lindridge 1850.

[76] Robert 1841: 174.

[77] Podmore 1906: 516–17; Barker 1840a: 1; *Quarterly Review*, 65 (1840), 488, 519; *Tait's*, 7 (1840), 550.

socialism (giving it an 'accidental prominence' that could be found regrettable). Just occasionally an Owenite will say bravely that 'the Bishop has done us good', and the evidence is that favourable interest in socialism among the working class actually increased considerably in 1840. But among Owenites themselves the sense is oddly strong of a movement at risk, successfully hounded by enemies who have harped on the sexual theme.[78]

The anti-Owenite press, at its most strident, would have it that 'hissing was general' when a socialist speaker tried to defend divorce, and even that 'loud laughter' greeted the notion that a couple should not continue to live together in a state of misery. This is to strain credulity. It is unthinkable that such ideas would not have been listened to with interest, sometimes painful interest, by working-class audiences, despite the undoubted moralism of working-class culture. The opportunity for legal divorce created almost twenty years later by the Act of 1857 was taken advantage of to a surprising extent by working-class couples (out of a sample of over 300 of the petitions filed in the first decade of the Divorce Court's operations about a quarter were from working-class individuals).[79] But if the working-class code did not exclude transgressing rigid notions of marriage in the interests of happier and closer unions, the sexual free-for-all of 'promiscuous intercourse', so often alleged to be an Owenite ideal, may have been genuinely repugnant to working-class audiences. On grounds of self-esteem there may also have been genuine anger, as was frequently claimed, when anti-socialists quoted at their listeners Owen's declaration that 'almost all' marriages were a kind of prostitution. These do sound like serious 'thorn-bushes'. One potential follower of Owen at least was alarmed by the allegation that 'you generally profess a kind of latitudinarian principle, with regard to the sacred institution of marriage' and wanted this repudiated. One very important phenomenon tended to justify fears that allegations about the Owenite sexual code were doing damage to the cause: namely, the reaction of working-class women.

Women were conspicuous attenders at Owenite meetings, making up a fifth of the audience at ordinary lectures and even, it was said, a majority when the marriage question was featuring in a debate-series. As with the male component, these women seem to have been

[78] Sturmer 1840: 4; Grant 1841: ii. 185; Royle 1974: 50, 62–3, 67; *New Moral World*, 7 (1840), 1186; B. Taylor 1983: 139.

[79] *Anti-Socialist Gazette* (1840–1), 15; Savage 1992.

mainly from the upper levels of the working class.[80] It is clear that their anxieties about the socialist sexual system were deliberately played on by anti-socialist propagandists—with 'Women of Bath' or 'Women of Bradford' or just 'English women' at large being more explicitly targeted, in spoken and printed propaganda, than any other potential working-class ally against the 'farmyard morality'.[81] There is much evidence that these warnings were astutely directed. Robert Buchanan, one of the social missionaries, recalled that fears that Owen wanted to destroy the marriage system 'drove the wives and mothers of the toiling classes to absolute frenzy', and there are persisting glimpses of women made distrustful of Owenism and the community ideal. Women are reported as trying to walk out of meetings, as calling out 'you wish to make all prostitutes', and, at Newcastle, Paisley, and Whitehaven, as being aroused to literal frenzy against visiting and local socialists.[82]

It is notable that in two of these episodes of actually violent behaviour by groups of working-class women the victim was Margaret Reynolds, later Mrs Chappellsmith, who was a major agent of damage-limitation from 1839, loyally denying in the press and on the lecture platform that any real change in the existing marriage system was intended by socialism. At most she 'pointed out the certainty of unions contracted under the new arrangements . . . being permanent'. It is possible that she was not even acquainted with Owen's 1835 *Lectures*. At other venues she is reported as being warmly received by female audiences, as were a pair of male campaigners, George Fleming and George Simkins. These men and the Owenite leadership generally were energetically reassuring women that there were no intentions to change the marriage code in the foreseeable future—a theme which, as I have mentioned, goes back to the pages of the *Crisis* in the mid-1830s.[83]

The female reactions in Newcastle and elsewhere are a striking testimony to the strength of women's fears about a relaxing of the

[80] Letter from J. F. D. (1 Apr. 1832) in Owen Collection, Holyoake House, Manchester; *Anti-Socialist Gazette* (1841–2), 1; *Christian Lady's Magazine*, 13 (1840), 378; B. Taylor 1983: 57, 76; Royle 1974: 64; Bowes and Jones 1840*b*: 1; R. Owen and Legg 1839: 24.

[81] Campbell and Roberts 1838: 38–9; Owen and Brindley 1841*a*: 62; Robert Matthews 1840: 51–2; Scoresby 1840: 9; Bowes and Jones 1840*a*: 19; see also Blackburn 1840.

[82] H. Jay 1903: 6; *New Moral World*, 5 (1839), 23, 9 (1841), 402, 10 (1841), 48, 13 (1845), 319; *Robert Owen's Journal*, I (1851), 195; Dalton and Jones 1838: 41; Campbell and Bannister 1839: 82; *Anti-Socialist Gazette* (1841–2), 82–4; Taylor 1983: 189, 207.

[83] *New Moral World*, 5 (1839), 3–4, 23, 556–8, 7 (1840), 1187, 1307, 8 (1840), 252; Ainslie 1840: 5.

laws of matrimony. The suspicions and alarms about the implications of Owenite policy for their sex expressed by women within the movement, long before the external attacks, were in this sense better founded than the equivalent anxieties of male Owenites. Frances Morrison, in a remarkable tract based on a lecture to women in Manchester, was the only female Owenite publicly to embrace the principle of a new system of heterosexual unions.[84] It is a fair assumption that the main impetus to these fears, especially as exhibited by working-class women, was the prospect that more informal marriage rites and easier divorce would deprive wives of what little protection against male exploitation the existing laws offered. The Owenite scheme looked very like a charter for men to have all the pleasure and no liabilities.

Nevertheless, interestingly, the enemies of socialism did not always trust that self-interest would ensure that women rejected the Owenite programme. John Relly Beard, later a prominent Manchester Unitarian, believed that there was 'one half of society arrayed against it in mild but determined hostility', but he did *not* think it was women's adverse prospects under socialism which necessarily caused this attitude: 'even when their interests did not prompt them to opposition, their heart, their best affections rebelled against the idea'. He appears to acknowledge, in fact, that on a calculation of interest the socialist system could have some charms, because it 'must tend to make existing connexions irksome', and 'there are many, very many unhappy homes'—a remarkable concession to Robert Owen's own kind of account. It was possible, especially once the official version of Owenite policy had started to lay such stress on the stability and richness of Owenite marriage, to see considerable advantages for women in what was being proposed. A woman wrote to the *New Moral World* in 1841 acclaiming it as the only system 'which elevates her [woman] to her real dignity as a human being', and this may well have been the spirit in which the teachings of Fleming, Simkins, and Margaret Chappellsmith were welcomed, when they were listened to.[85]

The Owenite vision was by now scarcely adventurous. Except for the requirement that divorce should be available to ordinary citizens if it were available to the wealthy, there was little, apparently, to distinguish it from other kinds of ardent matrimonialism in the period.

[84] See above pp. 146–7, also *Crisis*, 2 (1833), 279–81; F. Morrison 1840: *passim*.
[85] Beard 1839: 150–1, 158, 160–6; *New Moral World*, 9 (1841), 212.

But it had arrived at this point by its own highly unusual route, as contrasting strains of libertarianism and high-mindedness, of pro-sensualism and anti-sensualism, proved not to be reconcilable. The striking unease about Owen's sexual policy shown by women was a matter of prudent foresight of consequences, but also, I believe, sprang from the anti-sensual side of the socialist mentality. Owenite feminists such as Anna Wheeler were in the anti-sensual tradition of Mary Wollstonecraft, which I will briefly sketch in the last pages of this study. As Owenites they were also in contact with a rather different tradition of this sort, in which progressive hopes for humanity were bound up with anti-sensualism very much in the spirit of the zealous asceticism of certain religious groups.

Robert Owen bought New Harmony from an existing communitarian movement called the Rappites. This sect practised sexual abstinence, though with less strictness than the Shakers. When Owen wrote to the leader, George Rapp, in 1820 enquiring about their operations, this aspect of Rappite doctrine had been reported in several accounts of the sect published in Britain. These had apparently reached working-class readers, and there is no doubt that for Owen and many other socialists the colonies being established in America by sects such as the Rappites and Shakers were exciting realizations of the ideal of co-operative, egalitarian life. When Owen reached America in 1824 one of his first ports of call was the Shaker community at Niskeyuna, and several leading American Owenites had links with Shakerism.[86] Owen believed that these American communities could have achieved a sharing of goods without also downgrading marriage, but otherwise his attitude to sectarian celibacy is not known. He may have agreed with his son's view: that it was not the same as temperance (which was desirable), and more like licentiousness turned on its head. The *New Moral World* often carried items on Shakerism; the Shaker doctrine of celibacy, when it is mentioned, is treated as an example of the aberrations of religion. But the example of the Shaker regime could be useful to socialist missionaries in rebutting charges that community life necessarily led to sexual promiscuity.[87]

One colourful communitarian of the 1840s, John Goodwyn

[86] Arndt 1965: 226–7, 269–71; National Community 1838: 7; J. F. C. Harrison 1969: 105–8; Bestor 1950: 46–7, 106.

[87] R. Owen 1830: 74; *New Moral World*, 11 (1843), 66, 13 (1845), 267–8; Dalton and Jones 1838: 48.

Barmby, who started as a Chartist and contributor to the *New Moral World*, did found an English sect which had severe sexual restraint as a leading principle. Barmby's 'Communist Church' established a community called the Concordium at Hanwell, Middlesex, and it is evident from its periodical, the *New Age*, that utopian goals and repression of sexual drives were taken to be intimately connected: 'the whole fall of man is involved in sexual coition . . . no great proficiency can be made in human progress to the universal good, until an effectual work of self-denial is begun here'. Barmby and his wife Catherine did not believe that abstinence should be carried, as with the Shakers, to the point of endangering reproduction: nevertheless, their goal was 'to copulate . . . not more than once in two or three years'.[88]

Barmby had links with celibacy-minded American groups, including the Shakers, and in England his closest association was with another 'Concordium', that established at Ham Common, Surrey to implement the teachings of the influential educationalist and seer James Pierrepont Greaves. Ham Common and Owenism in the early 1840s, or at least certain important figures in Owenism at this period, were intertwined. The celibacy so much encouraged at Ham Common was one aspect of a general anti-sensual Greavesian ethic, which included vegetarianism, and was bound up with the sect's Neoplatonic metaphysical scheme. (There was a distinct current of austerity, even of primness, in how some Owenites regarded the life of the body too: Owenite recreation could be a jolly but anaemic affair, with no alcohol, and dancing accepted because it produced 'a healthy tone of body and mind').[89] Thomas Frost was an Owenite who felt very drawn to Ham Common, but for whom the community's bar on sex was an unwelcome surprise and deterrent. He was told not to marry his fiancée before joining, but to bring her to live austerely with him in the community if he wished: marriage would have to wait on spiritual improvement, and even when performed should ideally not be consummated. The wife of Robert Buchanan, one of Owen's social missionaries, did join the Concordium; she left after a short time, perhaps because her married condition was deemed to exclude her from the 'inner, or perfectly sacred, circle'.[90]

---

[88] *New Age* (1845), 116–20, 200–2; B. Taylor 1983: 174–7.

[89] Armytage 1961: phase iii, ch. 1; J. F. C. Harrison 1969: 127–31; *New Moral World*, 11 (1843), 255.

[90] Frost 1880: 43–6; H. Jay 1903: 10.

Alexander Campbell, another social missionary and the most illustrious Owenite to be involved with Ham Common, was actually resident in the community from 1842 for about two years, and taught Greavesian doctrine while remaining an active socialist. Campbell met Greaves in 1838, just as his responsibilities as a missionary were taking him on to platforms to speak up for the high-mindedness of the Owenite sexual creed, and he was drawn into sympathy with Greaves's doctrines in a long exchange of letters. It is a fascinating token of the anti-sensual potential of the Owenite concept of marriage that Campbell could later officiate at a socialist civil wedding under the terms of the 1836 Act (Owen had set the example for this procedure), and give an address about 'sacred marriage', the spiritual growth of the couple, and their producing children 'not spoiled in the begetting'. This was evidently a promulgation of Greavesian doctrines of sexual abstinence, for one of the teachings at Ham Common was that of 'chaste generation', whereby sexual restraint by the married led to finer offspring. Owen's rhetoric about the purity of the new marriages had come home to roost in a surprising way.[91]

The sense of sectarian celibacy as an inverted form of libertarianism hangs around the community experiments, especially Barmby's Communist Church (which was alleged, though probably without any justice, to be a haven for sexual depravities).[92] In the strange career and writings of James Elishama 'Shepherd' Smith[93] actual oscillation between sexual austerity and eroticism is more overt. Significantly, Smith's first discipleship was as a circumcised follower of that ambiguous figure, John Wroe, in Ashton-under-Lyne. He arrived in London as a Southcottian preacher in 1832, but soon rose to a brief and surprising eminence in Owenism. He became a lecturer second only to Owen himself in importance, and editor of the movement's journal, the *Crisis*, for a year.

It appears that in the short period of his attachment to socialism, and immediately thereafter, Smith accepted Owen's teaching on the future of sexual unions, which was in its most liberal phase at this date. He left the socialist camp in 1834, and by the time of the 'momentous crisis' of 1840 was critical of both Owen and his

[91] Marwick 1964: 12–13; *National Reformer*, 15 (1870), 172; J. F. C. Harrison 1969: 128–9; *New Moral World*, 13 (1845), 70; Greaves 1843: i. 4, 51, 67; H. G. Wright 1840: 5; Armytage 1961: 176.

[92] *Antidote* (1842), 15.

[93] Best explored by J. F. C. Harrison 1969: 112–22; see also Saville 1971; B. Taylor 1983: 167–72; McCalman 1992.

enemies: the former was guilty of 'sensual' immoralities, while the Bishop of Exeter was guilty of 'spiritual' ones.[94] But this should not be taken to imply that sexuality or even eroticism had dropped out of Smith's creed. The bizarre religion of 'Universalism', which he started to promulgate in his periodical, the *Shepherd*, in the latter 1830s, stemmed from Southcottism, from Swedenborgianism, and from Saint-Simonism; imagery of gender and sexual consummation is an essential feature of its cosmology and its historical scheme.

Above all, by remaining unremittingly hostile to the 'sensual' but at the same time projecting into the spiritually improved human state an era of sexual freedom which outdoes anything recommended by Owen, Smith keeps extremes of pro- and anti-sensualism in play in a unique fashion. 'It is a horrid thing love, at least such love as this world furnishes'. On the other hand, 'in heaven . . . men and women make love to each other just as they please', 'the two sexes unite in all the endearments of love'. Smith was perfectly aware of the strange contrasts in his deportment, indeed pleased by them:

I suppose I never once in my whole life gave utterance to a profane oath, and my speech has always been free from obscenity, immodesty, or indelicacy . . . I . . . have often thrown down Shakespeare because my feelings were irritated by it. I have even envied those that could read it. I have actually tried to overcome the weakness, and cannot; and yet . . . I have written letters to friends which, if discovered and published, would look almost . . . obscene— analogical letters to theologians, treating of the spiritual analogies of the whole human body.[95]

Such keeping of the sexual theme in play, in a now-you-see-it-now-you-don't game of denial and acknowledgement, is additionally erotic because Smith seems to have specialized in female audiences, perhaps in a rather calculating fashion. He became a purely commercial editor in 1843 when he founded the successful *Family Herald*, and the magazine dealt with requests from women for advice on their love-lives. Smith was patronized financially by two widows in the 1840s, and his correspondence with these and with the maritally unhappy Lady Lytton some years later, as printed in his nephew's memoir, makes very strange reading. The element of erotic teasing, with Smith's own chastity used louchely to arouse thoughts of its

---

[94] J. F. C. Harrison 1969: 120; Saville 1978; B. Taylor 1983: 168; *Shepherd*, 1 (1835), 274; J. E. Smith 1840: 11, 19.
[95] W. A. Smith 1892: 380, 395; J. E. Smith 1833*b*: 133.

violation, is sometimes quite overt. He recalls to Lady Lytton how, as a young man,

unless God created a woman on purpose for me, I felt that marriage was not for me. But I have all my life entertained the most exalted ideas of Love . . . though the ladies know nothing of me personally . . . yet it seems that my spirit has some attraction for them . . . Do not, however, let any wicked thoughts enter into your head . . . It is all in spirit . . . One of the greatest blessings of heaven is love, and it is only there that men and women will thoroughly enjoy one another.[96]

In leaving the story of the Owenite would-be sexual revolution it is worth balancing this group of individuals, for whom the socialist retreat from a radical policy was evidently not at all uncongenial, with an Owenite who helped to cover the retreat but who later declared his disappointment with it. He too was probably not alone in his reaction to the events of 1838–42. Charles Southwell, a social missionary and a notable free-thought campaigner,[97] in 1840 penned two pamphlets and a statement in the *New Moral World* which offered standard conciliatory arguments on behalf of Owenite sexual policy: it would not lead to promiscuous intercourse (women are not so depraved); some 'silly friends' had misinterpreted it (the example of the Shakers and Rappites was more to the point); it was at present only pleading for divorce, and whatever form of marriage was introduced in due course (and there must be some) would 'give a high moral tone to society'.[98]

Southwell was a robust personality, however, and the more significant of his two pamphlets, *An Essay on Marriage*, is enterprising in its development of Owenite sexual doctrine. The effect is to expose, as no other socialist does, difficult issues in Owen's determinist approach to sexual reform. In effect Southwell distinguishes two deterministically controlled aspects of our sexuality. There is the sexual 'instinct', which exists to some extent in every individual, driven by our biology and enhanced or reduced by factors such as 'climate, soil, etc', and which cannot be denied without serious damage ('like steam . . . it will force its way'). And there is 'passion', which is learnt, and controlled by the values of a particular culture (here Southwell accepted the doctrines of Rousseau's *Emile*). This analysis has quite

---

[96] W. A. Smith 1892: 337.

[97] For his career see Royle 1974: 69–72 and *passim*; *National Secular Society's Almanack* (1870), 38; F. B. Smith 1967.

[98] *New Moral World*, 8 (1840), 340–1; Southwell 1840a: 13, 16, 1840b: 1–3, 6.

complex implications for sexual reform. According to Southwell, any legislation must be geared to the standards of the society, to what counts as damaging sexual behaviour in that community. He is thus led into very tolerant judgements on, for example, prostitution in the classical world and oriental polygamy.

Southwell is nevertheless able to emerge with considerable consistency as a proponent of monogamy for English society, and a politically progressive proponent at that. In England, he says, 'a plurality of wives, would . . . lead to the disorganisation of society'; existing transgressions of this code are due to inequalities of wealth, with the rich 'maddened by luxury, filled with fiery juices' preying on working-class girls. The true cultural environment in the nation is other than this: 'there is a deeply felt want, an aspiration after knowledge, a desire for change, an appreciation of the beautiful and art . . . all will acknowledge the supremacy and act under the guidance of the moral feelings'.[99]

Given how Southwell's argument comes out in 1840, it is on the face of it surprising that, in his highly readable *Confessions of a Free-Thinker* of a few years later, he is bitterly critical of socialism's trimming of its reforming policies on sex in response to conservative outcry. But Southwell never lost sight of his dual model of human sexuality, and he probably felt that Owenites around 1840 had gone too far towards conceding what he himself had never allowed, 'an arbitrary moral standard' of sexual restraint, universally applicable. It is characteristic of his sense of the interplay of irremovable sexual drives and locally varying sexual codes that in the *Confessions* he deplores, like Rousseau, the artificial 'precocious development' of the sexual instinct in boys in modern urban life, but favours the very unRousseauesque solution of well-supervised brothels.[100] His remarkable sexual reminiscences in chapters 5 to 8 give details of convulsive interactions in his own life between drives and cultural conditioning: especially in the passages on his tormented marriage to an unfaithful wife, near-bigamy with her good-hearted aunt, and remorseful return to the adulteress. Southwell's *Confessions* can in fact be recommended as perhaps the most unjustly forgotten memoir of the period.

[99] Southwell 1840*b*: 5–6, 13–14, 17–18, 1845: 19.
[100] Southwell 1845: 19–22, 60; 1840*b*: 6.

Chapter Four

FREE-THOUGHT VERSUS FREE LOVE

I

A New Climate

THERE were other radical movements in the 1830s and 1840s, in addition to Owenism, whose programmes contained a sexually subversive potential. But these are shadowy affairs, certainly much smaller in scale than Owenism, and the indications are that their adherents were nervous about the libertarian doctrines they were in some sense saddled with, very much in the fashion of most socialists. Both the Saint-Simonian and Fourierist movements in England were importing from France programmes of social renovation in which greater sexual promiscuity was contemplated. Whatever the truth of the lurid stories of Saint-Simonian activities at Mile End, it is certain that the official English publicists for the movement were extremely sensitive to allegations about 'community of women', and like many socialists they responded by trying to steal the high moral ground on sex from their enemies. In the background, as with Owenism, there seems to have been a real division of attitude among Saint-Simonians on the liberalizing of sexual codes.[1]

The ethos and plans of Fourierist communitarianism, or 'pha-lansterism', were particularly congenial to Owenite thinking, and the

---

[1] Hancock 1837: 35, 43; 'Fontana and Prati' 1833: 22–8; B. Taylor 1983: 467; Pankhurst 1957: 108.

numerous treatments of the subject in the Owenite press, especially
the *New Moral World*, must represent the largest exposure given to
Fourier's thought in this country. The sexually transgressive element
in his programme, the fact that 'the cohabitation of the sexes will not
be prevented', does emerge in this coverage, but the general tenor of
the comment is none the less firmly to deny it and, again, to claim the
high ground, even that of 'christian purity', for Fourierism. The
prominent translators and promoters of Fourier were also eager to
rebut the suggestion that his brand of freedom included a relaxing of
sexual arrangements, but this was not easy to square with the plain
evidence of his teaching. The leading English Fourierist, Hugh
Doherty, actually went to the point of criticizing the master for his
ideas on sexual unions: 'Fourier's disciplines of love are not in keeping with his views of liberty'. Among consenting Fourierists in private, of course, things may have been different: it is usually said that
the cohabitation of George Henry Lewes's wife with Thornton Hunt
was attended by phalansterian beliefs.[2]

Liberationist activity ceases around the mid-century, and then
reawakens in the early 1860s as an aspect of the secularist movement,
to intensify gradually over two decades. In other words, it moves to a
rhythm often noticed in Victorian developments, of two episodes of
turbulence and controversy flanking an episode of stability and consensus, the years of the Great Exhibition. One could be forgiven for
supposing that in the case of sexual liberationism and its struggles
there is something of a monotonous repetition of old patterns. In
1862 the secularist lecturer Harriet Law was attacked by Reverend
Clarke in a pair of speeches delivered in Leeds: he accused her of
seeking to 'indoctrinate her sex' in the depraved 'principles and practices he had just quoted'. Clarke had been citing the anonymous volume of sexual medicine, fertility theory, and free-love propaganda,
The Elements of Social Science, first issued in 1854 or 1855. Mrs Law
(as sympathetically reported in the free-thought journal, the *Secular
World*, edited by George Jacob Holyoake) vehemently denied that
she agreed with the views of the *Elements*: 'she had again and again
repudiated the book'.[3] But it was advertised and several times recommended in another free-thought periodical, Charles Bradlaugh's

---

[2] Bestor 1950: 89; *New Moral World*, 2 (1836), 58, 66, 88, 9 (1841), 210, 10 (1842), 175, 11
(1843), 27; Gamond 1841; 121; *People's Journal*, 3 (1847), 262–4, 283–4, 345; Fourier 1851: i.
xxx.

[3] *Secular World* (7 June 1862), 57.

National Reformer, and the author of the *Elements*, George Drysdale, was a frequent contributor to the latter under the alias 'G.R.'

This is all certainly very reminiscent of the tribulations of Owenite sexual policy and its presentation, over two decades earlier. There were important continuities of personnel also. Holyoake had started his radical career as a socialist missionary. One of the most insistent voices attacking Bradlaugh's endorsement of the *Elements* from within the secularist party was that of William Stewart Ross, or 'Saladin': Ross had been a protégé of the veteran Owenite anti-sensualist, Alexander Campbell.[4] The religious or quasi-religious viewpoint of Campbell and 'Shepherd' Smith has a most conspicuous descendant in the theosophical creed of Annie Besant, embraced by her in 1889, and bringing with it a complete repudiation of the neo-Malthusian policies—that is to say, birth-control policies—which she had championed brilliantly for a dozen years.

But a shift has taken place. In the strange, unsympathetic pages in her *Autobiography* concerning her rejection of neo-Malthusianism at the behest of Madame Blavatasky, Annie Besant does not claim a spurious continuity in her experience. It was a matter of discrete visions: one (the 'Materialist') is of course lower than the other (the 'Occultist'), but on its own terms perfectly respectable, particularly in its concern for human suffering in the here and now. There had thus been real difficulty in accepting Blavatasky's ruling on birth-control work:

I felt that she was right, and though I shrank from the decision—for my heart somewhat failed me at withdrawing from the knowledge of the poor, so far as I could, a temporary palliative of evils which too often wreck their lives and bring many to an early grave . . . yet the decision was made. I refused to reprint the 'Law of Population', or to sell the copyright.[5]

In highmindedness the moral code which had accompanied Annie's neo-Malthusianism did not yield anything to theosophy, we are told: it had contained 'nothing which for one moment implies approval of licentiousness, profligacy, unbridled self-indulgence'. The record of Annie Besant's views at the time when she was still an unreconstructed neo-Malthusian, in the late 1870s and 1880s, supports this description—but only up to a point. Her great birth-control pamphlet just referred to, *The Law of Population* (probably first issued in 1879), is calm, if in a moralistic way, about the prospect of a

---

[4] Hithersay and Ernest 1887: 13–16.   [5] Besant 1893: 243.

certain amount of 'licentiousness' as a price worth paying for the great general good of population control: 'are all to suffer, so that one or two, already corrupt in heart, may be preserved from becoming corrupt in act?' That touchstone book, *The Elements of Social Science*, is here recommended in general terms for its neo-Malthusian arguments—although Annie tends increasingly, in the years up to her conversion, to temper her approval with an emphasis on how 'I . . . dissent utterly from its view on the marital relation' (and she also attributes to Bradlaugh a dislike for the liberationist creed of the *Elements* which, as I shall show, he may not have felt).[6]

In other words, there is missing from the mentality of this major defector from sexual reform in the 1870s that austerity which was so often part of the very fabric of Owenite sexual thought, always threatening to subvert its radicalism. Similar remarks apply to George Jacob Holyoake, though in his case the tokens of a new receptiveness to liberal sexual policies are obscured by a chronic tendency to fudge and waver. Holyoake is altogether a puzzling mixture, for he was also the doughty and revered founder of the secularist movement in Victorian England, steering in this direction the energies of the Owenite co-operative ideal, and of stagnant radicalism generally, in the late 1840s and 1850s. Always the advocate of a conciliatory, refined propaganda for free-thought, he was supplanted as the leader of secularism around 1860 by Charles Bradlaugh, an altogether more robust, aggressive, and, by comparison, coarse presence, though Holyoake remained the hero of a moderate party within the secularist movement which had several clashes with Bradlaugh.[7]

Sexual policy was often the trigger, or at least the pretext, for these internecine quarrels, with the *Elements of Social Science* again playing an important role as a focus for conflict. Holyoake, as editor of the *Reasoner*, wrote an early review of the *Elements* in 1855 which praised in particular its provision of reliable birth-control advice. He accepted Drysdale's argument that both celibacy and overpopulation are 'dangers to society' which birth-control skills can combat, and he did not object to Drysdale's astonishingly bold version of the non-celibate life, in which all liaisons between affectionate consenting males and females after puberty are encouraged, including the most temporary. There was a recognition that writing about these questions is inevitably a ticklish matter ('in some respects all books of this

---

⁶ Besant 239, 1879: 39, 1881: 10–11, 1885: 100–23; *Daily Telegraph* (14 Nov. 1889), 5.

⁷ Tholfsen 1976: 264; Budd 1977: 26–9; Royle 1974: *passim*, esp. pp. 72–206, 275–85.

class are evils'), but the *Elements* is judged to have emerged with
more dignity than any previous effort. Holyoake does not, at this
stage, argue that a secularist treatment of these topics should, for the
sake of the movement, be especially cautious in its terms and views.

Later in the same year Holyoake endorsed the extreme liberalism
of the *Elements*, in the sense that he accepted the book as relevant to
the 'great evil' of orthodox marriage, though it is significant that he
now identified certain other documents from within free-thought
(such as Robert Owen's lectures on marriage) as being 'unhappy' and
'wanting in taste' in their expression, and thus frustrating an otherwise 'honourable' intention. Holyoake is here offering a mild correc-
tive to a series of letters in the *Reasoner* by Francis Newman which
were very hostile to the *Elements*, but looking back on these events
many years later he claimed that Newman's attack had enlightened
him as to the pernicious tendency of the book, particularly in
pointing out that 'the author regarded seduction as a physiological
necessity'.

As will become clear when I describe the *Elements* more fully, no
one who had examined it even perfunctorily could fail to come across
Drysdale's 'law of exercise' for the sexual organs, and in these later
accounts Holyoake is forced to claim that he had reviewed the book
favourably in 1855 'without looking carefully into his pages'—which
does not square with the evidence of the review itself. He also recalls
writing to Drysdale (indirectly, one must assume) urging him to
change the title, which at this stage was the later subtitle, 'Physical,
Sexual, and Natural Religion' (a phrase 'repulsive' for its notion of a
sexual religion, Holyoake felt). But Drysdale himself in his preface
writes of the change of title as if it were entirely his idea. Holyoake
may have been misremembering what did certainly occur, that he
obtained permission to add a subtitle in advertisements for the book
in the *Reasoner* (this subtitle read: 'Laws of Social Health, or
Dangers to Society from Celibacy and Over-Population').

Holyoake is alleged to have protested, with other secularists, when
the free-thought periodical the *Investigator*, first under the editorship
of W. H. Johnson and then that of Bradlaugh, paid some modest
attention to the *Elements* in the late 1850s. He returned briefly to
centre-stage in the secular movement as co-editor of the *National
Reformer* with Bradlaugh in late 1861, in the wake of a schism
between Bradlaugh and his first co-editor, Joseph Barker. The main
bone of contention between Bradlaugh and Barker—who had been

the fiercest of the anti-Owenites twenty years earlier—was, yet again, the *Elements of Social Science* (Barker's hatred of the book indicates a further strand in the story of how libertarian sexual doctrines were received in the free-thought community, to which I shall return).

Holyoake therefore joined the *National Reformer* as a venerable figure displaying solidarity with Bradlaugh against Barker, the erstwhile preacher and scourge of the socialists (and he soon attacked Barker with great savagery under the pseudonym 'Quasimodo'). But Holyoake was also much concerned not to be identified with Bradlaugh's supposed advocacy of the *Elements*: he announced that he had the latter's assurance in 'conversation' that he was no disciple of Drysdale. Holyoake himself now declared the Drysdalean law of exercise, 'which makes health the measure of morality', to be 'utterly pernicious' (though the 'good intention' of the book, and its 'many neglected truths', were conceded). The issue was sufficiently a raw one with him that he resigned from the *National Reformer* less than six months later when he thought he detected his co-editor recommending sexual intercourse to a correspondent on health grounds (Holyoake thus had contradictory impressions of where Bradlaugh stood, and this is another question I take up later).[8]

On one reading Holyoake simply became increasingly moralistic on sexual questions over the years, but the truth seems to be more complicated than this. He was apparently a genuine believer in birth control, happy to defend publicly a broad neo-Malthusianism against the attacks of both Christians and fellow radicals. He may not have wished any associated loosening of sexual codes to go beyond a liberalized system of marriage, but he was no ascetic. At least, he expected his readers to join him in laughter against James Pierrepont Greaves and his disciples ('They preached abstinence from marriage, and most things else. Their cardinal doctrine was that happiness was wrong'). But there was something in Holyoake's mentality which made it difficult for him actually to stomach a developed, black-and-white statement of the implications of his liberal views, and he persistently contradicted himself on what would be an acceptable vehicle for sexual reformist ideas.

The best known of these contradictions involves the fact of Holyoake's being for some twenty years the British distributor of an

⁸ *Reasoner*, 18 (1855), 198, 19 (1855), 121–2, 22 (1857), 78; *Present Day*, 3 (1886), 77; G. J. Holyoake 1896: 30–1; Mackay 1888: 27–8; *National Reformer* (23 Nov. 1861), 5; *Secular World* (10 May 1862), 1.

American birth-control tract, Charles Knowlton's *Fruits of Philosophy*, which, a few years after he relinquished his interest, and in a new edition, was the occasion of the famous prosecution for obscene libel of Charles Bradlaugh and Annie Besant in 1877. Holyoake begged the couple to refrain from any mention in the courts of himself or of his late brother and business-partner, Austin Holyoake. When they declined, he wrote letters to the national press repudiating Knowlton's book, which could have damaged their defence. (It compounds the complications of Holyoake's position that Austin was himself the author of a booklet advocating birth control, *Large or Small Families?*, first published in the latter 1860s: this pays homage to the *Elements of Social Science*, and it is unlikely that Austin assisted George Jacob in suppressing the sales of their joint property, the 'quack' Knowlton, although the latter claimed that he did.)

Unhappily, greed may have played some part in these inconsistencies of behaviour by Holyoake (Bradlaugh accused him, in obscure terms, of making money out of sales of the *Elements of Social Science*). But there is also a history of Holyoake's welcoming and repudiating free-thought statements on sexuality, disinterestedly, in a continuous, self-defeating search for an ideal text which would somehow be both bold and utterly pure. Knowlton's *Fruits of Philosophy* he regarded as at least an improvement on Richard Carlile's *Every Woman's Book*, which he claimed had damaged the secularist cause. In his 1855 review of the *Elements* Holyoake implies that Drysdale is, in turn, an improvement on all predecessors other than Robert Dale Owen's *Moral Physiology*—but the latter falls under his ban in 1879. In 1873 Holyoake locates his ideal of purity in a medical idiom: 'it is the disgrace of medical men that no college of physicians give proper information to the poor, in language which the pure might read; and not have them victims of the verbiage or pollution of quacks'. According to other utterances by Holyoake it was precisely the fact that, in the *Elements of Social Science*, 'the terms belonged alone to a medical book on prostitution', and 'the heads of chapters rendered it a purely medical book', that made it 'execrable—if meant for family reading', and rightly deterring to potential secularists when they found the book on display in free-thought centres.[9]

[9] Budd 1977: 22, 28; Nethercot 1961: 126; *National Reformer*, 5 (1864), 201, 29 (1877), 455; McCabe 1908: ii. 82; Bishopsgate Institute: Bradlaugh Papers Item 538; G. J. Holyoake 1873: 22, 1875: i. 231, 1896: 30–2; *Present Day*, 3 (1886), 77.

Another prominent opponent of secularist free-love doctrines was Francis Newman, brother of Cardinal Newman. Though a Unitarian he was a kind of honorary secularist also, in the sense that his version of religious faith was a formative influence on Holyoake's brand of anti-atheistical free-thought. I have already noted him taking a stronger line on the *Elements of Social Science* than Holyoake, in the *Reasoner*, and he evidently felt that this text was playing a key role in a loosening of sexual codes in the 1860s which he deplored. He was conservative on sexual issues to the point of publicly opposing any kind of fertility restriction.

Marie Stopes, however, owned a letter from Newman in which he admitted to an Irish birth-control propagandist that his previous 'simple revulsion of disgust' against contraception was not appropriate ('it has gradually broken upon me that too large and earnest a body of men hold doctrines such as you promulgate'): unfortunately the date of this letter is not apparent. It is interesting that Newman always granted that human sexual drives, both male and female, had to be coped with in his moral scheme. Sometimes he argues that these are actually not strong enough to warrant birth-control practices on Malthusian grounds. He also quite boldly makes a case for nocturnal emissions being the mechanism whereby nature enables young men to remain chaste (in explicit disagreement with Drysdale, who had a more spermatorrhoea-minded view of the phenomenon), and, most interestingly of all, writes of there being a 'like provision in women', for whom too chastity is 'anything but pleasant'.[10]

I have cited the views of Annie Besant, Holyoake, and Newman at some length to suggest that the moral climate was different at the time of the second wave of liberationist activity in nineteenth-century England. History was repeating itself, but in a new key, and radical aspirations were nearer to fulfilment than before: which is indeed very much the general proposal about recent history offered in George Eliot's *Middlemarch*, a novel set in the early 1830s and published in 1871-2. There was also a new environment for sexual reform, socially and intellectually. I have in the first part of this study offered evidence that the 1860s saw a considerable relaxing of codes, especially among young middle-class people. Divorce became statutorily legal in 1857: thus there had come about the one reform of the marriage system which even very abashed Owenites had continued

[10] Newman 1861, 1867, 1869: 23, 1871; Stopes 1922: 14.

to urge. Darwin published the *Origin of Species* in 1859, and evolutionism in this and other versions may have done something to loosen the grip of the old Enlightenment-derived idea that man was categorically an intellectual being—in contrast to the animal and non-intellectual—whose moral standards flowed from this fact.

The mid-century also saw the demise of Chartism, and there followed two decades or so of quiescence in working-class political radicalism. It is the general argument of this book that progressive thought and sexual liberationism were antagonists, or at least uneasy companions, in the nineteenth century, and the latter could thus be expected to prosper in the years between the dissolving of Chartist organizations in 1848 and the founding of the Labour Party in 1893. And this fact about the political climate had a specific importance for one of the great themes of later nineteenth-century sexual reform, namely, birth control. From the time of the first popular campaigns for birth control, in the 1820s, the general left-wing response to these ideas had been hostile. Individual radicals, and the radical press as it has been surveyed by modern historians, commonly denounced material such as Francis Place's handbills and Carlile's *Every Woman's Book*. There was a perceived resistance among working-class groups at this period to the case for birth control, which presumably stemmed in part from this ideological hostility.[11]

Left-wing distrust of contraceptive policies arose not just from a general apprehension that these might be a device for maintaining an inegalitarian status quo (with the working class, yet again, having the raw deal), but from the palpable connection of these policies with the loathed social doctrines of Malthus and other political economists. Although Malthus himself explicitly denounced contraception in the 1817 edition of the *Essay on Population*, there are recorded instances of people misunderstanding his position, or at least thinking of the advocates of contraception as 'Malthusians', even at this early date. Francis Place put his name, unusually for him, to his defence of Malthus's demographic theories, the *Illustrations and Proofs* of 1822; the references to birth control in this volume are brief and general, and all Place's more specific published propaganda on the topic is anonymous, but he tirelessly promoted his contraceptive policies in private letters and conversation, finding his efforts often resisted

[11] James 1979: 386; Field 1931: 95–8, 116–18; Langer 1975; McLaren 1978: 69–77; Boner 1955: 101–3; Hollis 1970: 231–2; Himes 1930b: 311, 1936: 222; Ratcliffe and Chaloner 1977: 95; Prothero 1979: 205–6.

'because I am "a *Malthusian*!"'. These activities were even mentioned in the hostile middle-class press, and Place's appropriation of Malthusian demographic theory commented on.[12]

Place evidently believed that the truth and importance of the basic Malthusian analysis of population outweighed any damage the association with Malthus might do to the cause of birth control in working-class eyes. He was in general unabashed in his intellectual alliance with political economy as a set of principles ('as clearly demonstrated as that the squares of the two sides of a right angle are equal to the square of the hypotenuse'), and with political economists such as Ricardo and J. R. M'Culloch, although none of them repaid the compliment by endorsing birth control. M'Culloch explicitly repudiated it after Place had contrived to have his lectures issued with a neo-Malthusian colouring—thus provoking a particularly frenzied attack on birth control in the radical press.[13]

This whole tendency was taken further in the main neo-Malthusian initiatives of the 1860s and 1870s, as conducted by Bradlaugh and his associates. There was if anything a more elaborate and thoroughgoing commitment to political economy among these activists, especially in the case of George Drysdale, who would not grant that any political economist was other than 'on the side of the people' (Place had said that 'a political economist must be Republican'), and who several times attacked Ruskin's *Unto this Last* as vigorously as any right-wing commentator. The Bradlaughites had a genuine indignation about inequalities of wealth and power in England, but they were firmly opposed to a socialist analysis of these matters in terms of the antagonism of capital and labour, and to socialist solutions of expropriation. Drysdale was in favour of co-operatives, but not of socialist communities.

Bradlaugh was well aware that the overt Malthusianism of the birth-control policies he promoted from the early 1860s was a sticking-point for many working-class families, 'who have been taught to look upon the political economist as the poor man's foe'. His magazine, the *National Reformer*, often received letters from working-class correspondents dismayed or disgruntled that Bradlaugh and his colleagues were apparently playing the bosses' game in their sponsorship of contraception; working-class men also expostulated with

[12] James 1979: 375, 386; Trelawney 1831: i. 5; Ryan 1831: 45; Wilson 1831; Miles 1988: 150–1, 153–5, 270; Himes 1930*b*: 303, 311.

[13] Himes 1930*b*: 313, 317, 1929*b*; Miles 1988: 144.

him at meetings. Formal socialist opposition to birth control does not develop until the 1880s, however, when the embryo Labour Party consistently denounced the efforts of the Malthusian League as it sought to build on the great propaganda success of the 1877 trial, and thus 'caused the League more soul-searching and more frustration than any other hostile body of opinion'. The 'impossible chasm' between the Malthusian and the socialist analysis of poverty continued to make for antagonism well into the next century.[14]

If the decade of the 1860s was thus politically a favourable moment for a candidly anti-socialist and pro-Malthusian birth-control campaign to make headway, there was another feature of the climate of these years which was less propitious. National self-awareness had become more imperialistic, with the British Empire proudly regarded as an enduring and natural political entity. Emigration as a response to population pressures at home now seemed more congenial and logical than when it had been invoked as a somewhat desperate-sounding counter-argument to Malthusian arithmetic early in the century. Conversely, the Empire seemed to create a duty of fecundity for the Anglo-Saxon people who must occupy or administer it—so that venerable pre-Malthusian beliefs about the link between national power and population had a kind of revival.

Really strident appeals to Britain's imperial destiny, in the context of hostility to birth-control policies, are perhaps a feature of the next decade, but there was a long and much-noticed exchange of letters in the *Daily Telegraph* in July 1868 which suggests that the imperial perspective was very present to the minds of the nubile middle-class. Marie Stopes writes of this correspondence as if it were all about contraception, and there was some recognition at the time that this was an unstated issue in the series. But emigration looms larger and larger in these letters and eventually becomes their sole topic. They seem to be a good reflection of the preoccupations of the *Telegraph*'s readership, to judge by the many expressions of gratitude from the correspondents that the paper had opened its columns in this way.[15]

The currents of feeling on birth control which mattered, however, must in the last resort have been the affirmative ones, for there is the

---

[14] G. Drysdale 1868: 382, 329–30, 392–4; *Political Economist* (1856–7), 53–5, 100; Himes 1930b: 317; *National Reformer*, 1 (1861), 86, 223, 239, 335, (1 Nov. 1862), 2, (1 Aug. 1863), 7, (31 Oct. 1863), 2, 5 (1864), 414–15, 6 (1865), 19–21, 9 (1867), 410; Maclaren 1978: chs. 9 and 10; Bonner 1895: ii. 185–91; D'Arcy 1977; Searle 1983: 'Oxoniensis' 1883: p. i.

[15] Jenkins 1871; Bourne 1878; D'Arcy 1977; Stopes 1922: 8; *Daily Telegraph* (17 July 1868), 2; *Public Health*, 1 (1868), 208–11.

crude demographic fact of discernible fertility restriction from around 1860—surely the most important single factor to be aware of in comparing the Owenite period of liberationist activity with that inaugurated in the late 1850s. I have discussed this in practical terms in the preceding volume, and I wish now to review the ideology or culture of contraception, both in its recorded forms (chiefly the surviving printed propaganda) and in forms that have to be deduced and guessed at, with the thought particularly in mind that effective contraceptive practices were resorted to increasingly by English men and women in the second half of the nineteenth century. All historians of the subject would probably agree that the overt birth-control propaganda of the day had some responsibility for the spread of these practices, but it is arguable that this was none the less in essence an example of a venerable process whereby communities have controlled their numbers non-traumatically. On this account the barrier methods of birth control—though they constitute the great novelty of the nineteenth-century propaganda—do not explain why or how men and women started to restrict their fertility noticeably at this time. If, as I believe, the barrier methods *were* decisive, the propaganda on their behalf should be correspondingly important.

I have also in my last volume touched on one aspect of the ideological setting for nineteenth-century birth control, namely, the views and practices of doctors. With a few exceptions, and certainly in public, doctors were ignorant, or ignoring, of contraception, and violently antagonistic when they did acknowledge it. However one of the most celebrated of the published explanations of birth-control methods, George Drysdale's *Elements of Social Science*, is by a doctor. This work, first published in 1854 or 1855 and acclaimed by an admirer for 'having revived a subject which had become dormant from the close of the Socialist agitation in 1844',[16] is much more than a contraceptive manual: being an elaborately buttressed argument on behalf of promiscuous heterosexual liaisons. The intrinsic character of this extraordinary treatise—as I have perhaps already indicated—and the standing it achieved (especially as a result of Bradlaugh's championing of the book and its author through three decades of reproach and abuse) are sufficiently remarkable to deserve separate treatment as features of the sexual culture of the 1860s and 1870s.

<hr />

[16] A. Holyoake 1870: 1.

II

Contraceptive Culture

Many decades ago two American historians, James Alfred Field and Norman Himes, showed that the first campaign in Britain for birth control was to an unsuspected extent the work of one individual, Francis Place. He was proved to be the author of many hitherto unattributed items of propaganda, the instigator of others, and the man who had converted Richard Carlile to the contraceptive faith. The last of these discoveries gives an extra dimension to Place's influence, since it was Carlile's *Every Woman's Book* which stimulated Robert Dale Owen in American to write *Moral Physiology*, which in turn prompted Charles Knowlton's *Fruits of Philosophy*—the latter being the pair of booklets which were pre-eminent in the conventionally published literature of contraception throughout the first forty years of Victoria's reign. Francis Place's modern biographer, Dudley Miles, has since argued persuasively that his achievement went even further, by casting doubt on whether Bentham, James Mill, and Robert Owen—who had hitherto appeared to share with Place some of the honour of perceiving the utility of birth-control policies—reached such a point in their thinking independently.[17]

However, Place did himself say that he had come forward as the advocate of contraception in his *Illustrations and Proofs* 'with the concurrence of friends who were themselves afraid to encounter the certain obloquoy of such allusions'. It is clear that there was a group of Benthamite individuals who, whether spontaneously or under Place's influence, had become convinced of the urgent relevance of neo-Malthusianism, but who were also nervous of lending it public support. The most famous case is that of John Stuart Mill, at first active on behalf of Place both as the author of neo-Malthusian articles and in the distribution of leaflets and other literature, but who was so deterred by a brush with the law in the latter role, in the 1820s, that he never publicly advocated birth control again. Privately, however, Mill was still a believer, and his published utterances on population were in practice sufficiently ambiguous to permit a neo-Malthusian interpretation, which he did not repudiate. The story of his earlier active involvement in Place's campaign was never quite forgotten,

[17] Miles 1988: ch. 9.

and resurfaced after his death.[18] Other people in this group who supported birth control in varying degrees short of public espousal were Ricardo, William Ellis, both the Grotes, George and Harriet (the latter continued to lament for many years that no one would 'bell the cat' of birth control—Place's self-exposure apparently not being enough—and she is credited by G. M. Young with having done some small-scale propaganda), and probably Alexander Bain.

The rationalist freedom from moral taboos on which such intellectuals prided themselves was presumably one channel by which actual contraceptive practice entered middle-class behaviour. The 'Strong-minded Woman' of Horace Mayhew's 1853 satire espouses Women's Rights, 'conversaziones', Bloomer meetings, and progress—and 'it is extremely rare that the Strong-minded Woman is the mother of a family'. Several women came forward in secularist forums on behalf of neo-Malthusianism before Annie Besant (though they also complained that not many of their sex were prepared to join them: after 1877, at least in the leadership of the Malthusian League, the numerical contribution of women was virtually as great as that of men). The female attendance at early neo-Malthusian lectures was sometimes substantial, and it was said in 1869 that most thinking middle-class women had a belief in birth-control principles which they could confide to each other but which they were embarrassed to be more open about.[19] In the writings of male experts on population questions there is often a most eloquent silence on contraception when it seems inescapably relevant (a silence which excited the contempt of the *Saturday Review* in 1866), but just occasionally in the respectable periodical press there are affirmations of the rewards of fertility control for the marital experience of the middle classes. There is a letter in this vein in *The Times* of 1857 from a husband who signed himself 'W.A.L.'. The *Temple Bar* in 1869 paints a warm picture of sensual young marriage, unencumbered by children for 'a year or two'.[20] The Bradlaughite endorsement of contraception in the *National Reformer* was rather more effective in eliciting loyal declarations from birth-

[18] Himes 1929*a*, 1930*b*: 311; Bain 1882: 89; *National Reformer*, 29 (1877), 355; G. J. Holyoake 1873: 22; Lea 1881: 65–6.

[19] Miles 1988: 149–50; Grote 1862: 67, 78–9; Young 1960: 21; Clarke 1962: 180; Bonner 1895: ii. 16, 176; Field 1931: 124; *National Reformer*, 7 (1866), 289, 9 (1867), 345–6, 12 (1868), 92, 109, 141, 14 (1869), 196, 29 (1877), 283, 406, 24 (1874), 15; Horace Mayhew 1853.

[20] Fawcett 1871: 115; William Taylor 1874; C. Morrison 1854: 60; 'Seduction and Infanticide', *Saturday Review*, 22 (1866), 481–2; 'Young Husbands and Wives', *Temple Bar*, 26 (1869), 498–506; *The Times* (9 May 1857), 9.

controlling members of the public: one signed himself 'The Happy Father of a "Very Small Family"'. In a tenuous sense the hundreds of people who contributed to the defence fund for Charles Bradlaugh and Annie Besant in 1877—and who, to judge by the widely varying sums donated, came from all social levels—were militants on behalf of birth control (Annie Besant particularly liked to draw attention to the numbers of clergymen and their wives who had declared their support).

But the great majority of users of contraception must have been interested in the subject simply as it bore on their own lives, in a practical sense, with a relationship to the campaigning and publicity which was passive, dependent, and non-ideological. There is considerable anecdotal evidence of doctors being 'hard-pressed, by husband, and . . . by wives also' for advice on family limitation, and being made aware that 'most mothers never wish to have more than two or three children', with resulting 'persistent attempts . . . by ladies in good positions to restrict their families'.[21] Yet the signs are that doctors seldom obliged their patients on this score. What other sources of information did these lay men and women turn to? And in what numbers? And what lessons would they have learnt for their sexual and reproductive lives?

Place's propaganda was the first generally available advice on birth control ever to be issued in England. Unfortunately it is now not known with any accuracy how many people it might have reached. The central items were ingeniously conceived, being concise, well-printed handbills, of which there were probably three types in all, composing a logical series. Two consisted of single pages, headed 'To the Married of both Sexes' and 'To the Married of both Sexes in Genteel Life', carrying about 800 words on one side of a sheet some two-thirds the size of modern A4. The third type was a small four-page leaflet with a similar amount of text, entitled 'To the Married of both Sexes of the Working People'. The total number distributed of these handbills may have run into tens of thousands. They were ephemeral in the fullest sense, of course, and the operation to distribute them seems to have started in 1823 and stopped in 1826. There is also some reason to think that the most active work occurred in the north of England and Scotland.[22]

---

[21] *National Reformer*, 13 (1869), 235–6, 29 (1877), 266, 355; Goodell 1879: 432; Stopes 1922: 287.

[22] Miles 1988: 150–5.

The rest of the Placean propaganda of this period is by way of general advocacy of fertility restriction and explanation of Malthusian theory. But the handbills offer detailed contraceptive advice, with no more theory than can be worked into straightforward appeals to the experience of men and women in the various audiences addressed: the genteel are urged to think how their health and standard of living might be protected by control over reproduction, and the working people to think how it might bring higher wages and shorter hours. The contraceptive technique recommended by Place (with a brief reference also to coitus interruptus) is the vaginal sponge.

The modern reader may make the mistake of supposing that this choice of technique—nowadays so unfamiliar and thus ineffectual-sounding—shows Place to have been indifferent to practical aspects of contraception. The opposite is the case. The story of Place's being stirred into active birth-control campaigning by Robert Owen's showing him 'several pieces of sponge' from France rings true, for he evidently saw that this device had the great practical merit of being available to the ordinary couple. Place's handbills in fact belong to the literature of self-treatment, and their technical advice is by way of information on how to *make* a contraceptive out of sponge and 'a bobbin or penny ribbon', or 'double twisted thread' (here speaks the ex-tailor). In a manuscript draft of a fourth handbill Place mentions as an alternative to the sponge a tampon of 'lint, fine wool, cotton, flax, or *what may be at hand*' (my italics). Norman Himes observes that 'the condom is conspicuously absent' from these writings, but Place was surely thinking that to recommend a relatively hard-to-obtain device was worthless counsel. It should be mentioned, incidentally, that the vaginal sponge remained a commercially available device until well into this century, and may be found advertised in trade catalogues preserved in the Stopes collection at the British Library.[23]

Richard Carlile's *Every Woman's Book* of 1826 (which first appeared as a periodical essay the year before) may with more justice be accused of indifference to practical possibilities and outcomes. Carlile also favours the sponge, but his treatment of the competing methods—which are understood to include the 'dildo' and intercourse with the partners lying on their sides—does not suggest that he shared Place's sense of an urgent need for a reliable means whereby couples might mitigate the sheer toll of bearing and rearing children. *Every Woman's Book*, though originally inspired by Place, is in fact the founding statement in an alternative tradition of birth-control ideology, the libertarian tradition as opposed to the Malthusian. It is not addressed to the 'married', unlike Place's hand-bills, and in the text marriage (of a suitably dissoluble sort) is regarded as no more than the most desirable form of sexual union. Carlile holds that if the ratio of the sexes in a particular society turned out to be unequal either polyandry or polygamy would be the appropriate system. The overriding need, for health and happiness, is that the 'ripeness' of our sexuality be 'duly exercised': 'none unmatched that desire to be matched is the maxim of morality'.

There is also a fully developed, not to say cranky, subversive sexual ideology in *Every Woman's Book*: even the cross and resurrection of Christian faith are translated, in somewhat Lawrentian fashion, into phallus and erection. Carlile's liberationist intervention on behalf of birth control was often said, by people like Holyoake, to have been regrettable, but the intriguing fact is that his book sold extremely well, and may have reached more readers than Place's handbills. In the 1828 edition Carlile claimed that in the first two years of its life it had sold 10,000 copies. It remained in print for at least a dozen years, and its widespread influence was still being deplored in 1836. Interestingly, Place helped in the distribution of *Every Woman's Book*, even though the evidence is that he thought it ill-judged as a piece of birth-control propaganda. His recent biographer speculates that Place needed the resulting revenue to help defray the cost of his own handbill venture. If this is correct, Carlile's free-love-minded tract was financing Place's neo-Malthusian proselytizing of married couples.[24]

A miner who wrote supportively to Bradlaugh and Besant in 1877 recalled how at the time of his marriage he had heard about Robert Dale Owen's *Moral Physiology* and Charles Knowlton's *Fruits of Philosophy*, and 'made a purchase of both books and read them as carefully as if my life depended on them'. Thanks to the message of birth control ('what a blessing it has been to me and my wife!'), he had been able to have just three children over a period of fifteen years.[25] This would have warmed the heart of Francis Place as an example of precisely the enhancement of working-class family life which he looked for from a knowledge of contraceptive techniques,

[24] Carlile 1826: 9–10, 17–18, 37–9, 43; Hancock 1836: 12; Miles 1988: 153.
[25] *National Reformer*, 29 (1877), 307.

but in both the booklets read by the miner the case for birth control is rooted much more in a belief in the simple desirability of decoupling sex and reproduction than in Malthusian arithmetic. I have already described Owen's general position, which was confidently if equably hedonistic and pro-sensual. Knowlton's stance was very similar, though more self-consciously associated with a materialist metaphysics. This is the 'philosophy' referred to in his title (it is not clear if there is a play on 'fruits', incidentally), and his tract opens with a 'Philosophical Proem', which perhaps did not detain very many ordinary readers.

Owen recommends coitus interruptus, and is more or less disparaging of all other methods, including the sheath. Knowlton recommends the vaginal douche, and rightly claims as innovative the important proposal that douching could be used for contraception. (I have suggested previously that ordinary washing of the genitalia after intercourse was commonly performed by women in the hope of frustrating conception.) The actual equipment, a 'female syringe', 'may be had at the shop of an apothecary for a shilling or less' (Knowlton also suggests various spermicidal solutions which can be made up at home: though less of a Malthusian and more of a Carlilean than Place he is close to Place in his concern for feasibility).[26]

Knowlton is even more dismissive of the sheath than was Owen. With *Moral Physiology* and *Fruits of Philosophy* remaining in print for forty years, and being joined by other prominent birth-control tracts—first of all Drysdale's *Elements* and then, rather later, J. H. Palmer's *Individual, Family and National Poverty* (1875) and Annie Besant's *Law of Population*—which all in some measure discommend the sheath too, the pressure of opinion against this particular contraceptive method, at this level of the propaganda at least, is considerable. Sheaths of a modern type were being industrially manufactured by the mid-century and were quite cheap. It seems strange to our ideas that this device, which we treat as the basic instrument of birth control, should have had such an apparently modest status in the contraceptive culture of the last century.

However, as with the converse examples of the vaginal sponge (which remains the preferred option in Drysdale and Besant) and Knowlton's douche, we must not suppose that effective contraception was not achieved by new technical methods, just because nineteenth-

<hr>

[26] Knowlton 1877: 39.

century choices in this area were not ours. Moreover, occlusive vaginal devices of a more modern type than the sponge (devices for blocking the cervix) were probably also in use (though these are not mentioned in the advice literature until later editions of Besant's *Law of Population*: in other words, there is a clear likelihood—important also in relation to the sheath—that recommendations and choices about contraceptive methods were being made through channels other than conventionally published books and pamphlets, and I explore this shortly).

Nor should we be hasty in judging what the low status of the sheath implied about attitudes towards the sexual act, and the respective roles of heterosexual partners, though it no doubt tells us something about these matters. Knowlton makes one of the considerations in favour of the douche the fact that it is 'in the hands of the female': while the sheath is judged the worst option. On the other hand coitus interruptus he thinks preferable to the vaginal sponge. Robert Dale Owen is perfectly aware that coitus interruptus, which is his recommended method, puts the onus on the male, and happy with this fact.[27] In the advice literature there is a tendency, on balance, to endorse techniques which make women responsible for contraception, but the reason for this, to judge from the continuing emphasis on coitus interruptus, is not that men are generally considered exempt from such responsibility, or more entitled to unrestricted sexual pleasure than women (in fact, as I shall illustrate shortly, there are signs that the hedonistic, libertine male was expected to choose the sheath: George Drysdale did argue that a man's pleasure would be interfered with by the condom, but on psychological, not sensory grounds).[28] And cutting across whatever influence was exerted by prejudices of this sort was the effect of honest scientific calculation about the reliability of different methods: particularly the douche, rhythm, and sponge techniques (though this possibility will be denied by those who hold the fashionable and preposterous view that an individual can never make a scientific judgement which is not really a judgement of value).

Fruits of Philosophy sold about 1,000 copies a year in Britain over forty-odd years, and Owen's *Moral Physiology* perhaps 800 copies a year over much the same period.[29] To judge by the fact that sales of the former went up by something like fifty-fold at the time of the

---

[27] Ibid. 38–9; R. D. Owen 1877: 46–8.   [28] G. Drysdale 1868: 135, 350.
[29] Himes 1930a, 1936: 231.

Bradlaugh-Besant prosecution these figures fell far short of the potential demand for the books. Both booklets were issued by a series of secularist publishers and publicized mainly in the secularist press, though they were also advertised by booksellers in the lower-toned comic weeklies such as the *Penny Satirist*, the *Fly*, and Renton Nicholson's *Town*. Without these means of reaching at least a small part of their possible readership it is likely that they would have fared even worse. Four other pamphlets advocating birth control were published in the years 1840 and 1870 which do not seem to have achieved any circulation at all.[30] The struggle which such morally suspect writings had to reach an audience is partly reflected in how few copies of even the most successful publications have made their way into libraries: Carlile's *Every Woman's Book*, once a best-seller, is an exceptionally rare book, with only three examples on public deposit in Britain.

But the serious, disinterested tract on behalf of birth control is not the whole story of published contraceptive advice in the period. No less an observer than George Drysdale was surprised to discover, in the mid-1850s, 'how many cheap treatises and tracts have already been written on the subject of preventive intercourse'. Public curiosity about birth control was also being appeased by a literature which did not even aspire to book status, and which has now disappeared almost completely. A survey in Manchester in 1835 lumped together *Moral Physiology*, *Fruits of Philosophy*, and *Bridal Gem* (the three titles allegedly achieving sales of 600 copies a year in the city). *Bridal Gem* has vanished beyond recall; it was evidently a birth-control pamphlet in the same category as other offerings whose names can be gleaned from advertisements in the press. Examples are *Marriage for One Shilling*, *The Connubial Guide*, and two pamphlets by 'Charles Combe', *A Private Hint to Young Married People* and *Population Remedied and Rectified*. Many other ephemeral items of sexual advice are recorded, but, as one historian of birth control has warned, these must be treated very cautiously. It would be a mistake to assume that any of these publications referred to birth control except where there is a clear indication to this effect. Sexual debility, in particular, was a dominant, indeed overriding, theme in this

---

[30] 'Anti-Marcus', *Notes on the Population Question* (London, 1841); *Valuable Hints* (London, 1866); 'Oedipus', *The Marriage Problem* (Dublin, 1868); *The Power and Duty of Parents to Limit the Number of their Children* (London, 1868).

literature, and it never seems to be dealt with alongside birth control.[31]

To judge by the scanty evidence, popular birth-control publicity moves from an association with the sexual underworld to an association with the commercial manufacture of barrier devices. The major booksellers for this literature in the 1830s were 'John Wilson' and 'Henry Smith', who were basically Holywell Street purveyors of *risqué* works. By the 1870s (but before the Knowlton trial) the material took much more the form of advertising for manufactured products, often resourcefully disseminated in ways that attracted indignant notice: literature was pressed on married middle-class women by a 'respectably-dressed woman' in the Tottenham Court Road, and mailed to every girl seeking a post as a barmaid in the *Morning Advertiser*.[32]

What impact might this changing sub-literature of birth control have had on the practices of men and women? I have made the point in the first part of this study, in relation to the vigorous advertising which followed the Bradlaugh-Besant trial as recorded by Ethel Elderton, that the public response to contraceptive propaganda was not necessarily direct, in the sense that publicity for barrier devices could prompt resort to coitus interruptus. But in so far as the sub-literature was directly influential it would have encouraged the use of two methods, the sheath and the cervical pessary or cap, which the more disinterested tracts tended to disparage or ignore. The title *Marriage for One Shilling* makes it almost certain that this Henry Wilson pamphlet is about the sheath, called a 'protector' in the summary of contents. Henry Smith's semi-pornographic weekly the *Exquisite* has an interesting pair of articles in 1843 in which various contraceptive methods are canvassed, very much in the spirit of Owen, Knowlton, and Drysdale, but here the 'French letter' wins out over coitus interruptus, the douche, and the sponge.[33]

The *Medical Press and Circular* complained in 1870 about the extensive advertising for a certain Mr Orr's birth-control advice and products which he was securing in the daily press.[34] The examples given include an advertisement for a 'Pessarie Preventive'. Happily, a four-page catalogue for Orr's products, a most unusual survival, is preserved in the Wellcome Institute, pasted into an 1873 edition of

[31] *Political Economist* (1856–7), 86; Ashton 1934: 88, 141; F. B. Smith 1974; Peel 1964.

[32] *Lancet*, 1874(i), 357, 1875(ii), 650.   [33] *Exquisite*, 3 (1843), 108–9, 117–18.

[34] *Medical Press and Circular*, 1870(i), 460, 464.

the *Elements of Social Science*. Orr acted as agent for the birth-control products and publications of an American manufacturer, Robert Swift Bruce. The American-made pessary, he indicates, had been marketed by him in Britain since 1866 (he had also disposed of no less than 180,000 8½d. pamphlets of birth-control advice for women in the two years 1870–2).

Among other Bruce publications handled by Orr for the British market was a lecture on free love by the important American sexual liberationist Victoria Woodhull. In principle, there need not have been any link between contraception and libertarian sexual policies. Francis Place, though he privately welcomed the prospect of a time when 'the sexual intercourse should be free', was always careful to address his propaganda to the 'married'—and he surely felt that it was married couples who stood to benefit most from a knowledge of contraceptive technique.[35] Holyoake, and probably many others in the secularist movement, wanted birth control to be absolutely free from any damaging association with a relaxing of sexual codes.

In practice, such a separation never existed. Carlile's libertarian principles, rather than Place's Malthusian ones, prevailed in the Victorian years in the sense that both Owen and Knowlton are hedonistic in their approach, and are not interested in legitimating marriage. Indeed 'marriage', one may suspect, is often a euphemism in the sub-literature for every kind of liaison. 'Marriage for One shilling' could contain either the thought that early marriage is possible without any financial penalty other than the cost of a sheath, or that the unmarried may, for a small sum, enjoy the sexual satisfactions of married life without any fear of unwanted pregnancy. As I have stressed in the earlier part of this study, a fall in illegitimacy rates in the 1850s anticipated a decline in legitimate fertility: the implication is that birth control was widely resorted to in non-marital intercourse before it had won any substantial approval with married couples. The full range of meanings which contraception could bear—Malthusian and hedonistic alike—was insisted on, dominantly and inescapably, above all in one work: Drysdale's *Elements of Social Science*. And the privileged standing this book acquired in the secularist curriculum meant that free-thinkers were obliged to confront the libertarian potential of birth control, and give or withhold their assent.

[35] Miles 1988: 152.

III

George Drysdale and The Elements of Social Science

For Drysdale the demographic and the sexual problems of his society are twin visceral experiences of extraordinary intensity, two linked waking nightmares which he writes about in language like that of William Blake. With him the expression 'population principle' in fact denotes both these horrors: the horror of working-class poverty, which must haunt the consciousness of even the prosperous members of society, and the horror of sexual abstinence, the 'sexual disappointments and anxieties' which 'darken the whole sexual atmosphere' of the nation. What happiness and virtue people do achieve in this bleak world are, in the last resort, rendered a mockery by the savage struggle for food and love which is the underlying reality of a society at the mercy of the population principle:

> The second great corollary, that may be deduced from the principle of population, and which is enough to turn our hearts to stone, and to confound us with horror and bewilderment, is this; *that hitherto all happiness has been built on the misery of others* . . . This age, and all past ages in old countries, have been ages of *mutual destruction*. We eat the food of our fellow-beings, we breathe their air, we enjoy their loves, we suck their life's blood . . . the more deeply we consider this great subject, the more we shall perceive, that virtue, talent, happiness, have hitherto been mere delusions . . . perhaps the best thing you can do on the whole for your fellow-creatures, would be, to cease to live . . . Morality, medicine, religion, law, politics, are solemn farces played before the eyes of men . . . I wish to impress the deep conviction, that our present state of society is one so *horrible*, when we look beneath the surface, that it cannot possibly long continue as it is; that human affairs, when tested by the population principles, are found to be a hideous phantasmagoria, as if got up by some mocking fiend.[36]

The unmistakable altruism and earnestness of the author of the *Elements*, rising in passages like this to nothing short of anguish at the condition of his fellow beings, made a great impression even on the book's enemies. Hence their ambivalence was considerable, because they also found most of Drysdale's doctrines on sexuality exceptionally repugnant. His views were granted to be honest and 'consistent' even if 'abhorrent'; he was called 'earnestly, and even

---

[36] G. Drysdale 1868: 334–5, 338–40.

pathetically, philanthropic', and it was conceded that he achieved 'a certain Machiavellian insidiousness'.

There was in general a novelty in the conjunction of qualities offered by the *Elements* which made it understandably hard to categorize. It was remarked that the author's themes were like those of the worst sexual quacks, but he apparently had 'no meretricious professional object'; alternatively, he might have been 'writing ironically, and as a satirist, after the manner of a Rabelais or a Swift', were it not for his 'pure gravity'. Joseph Barker, in perhaps the most notorious of all attacks on the *Elements*, develops a strange trope to express this divided reaction. The author, declares Barker without the least show of evidence, 'is not an Englishman. It is a relief to know that this atrocious work is the production of a foreigner'; the book does contain 'many things . . . that are true and good', but 'they are nearly all taken from the works of Englishmen'. Thus everything morally alien in the *Elements* is coped with in the fiction of the 'nameless foreign quack' who is its author.[37]

There is a thin line of nineteenth-century free-love propaganda prior to Drysdale, but the *Elements* is on the whole marked off from these precursors precisely by its intellectual weight and elevation of tone. The politically radical libertarianism of William Thompson and the Spenceans, or of Richard Carlile and his disciples Hassell and Campion, who proselytized in the *Newgate Magazine*, is more consciously subversive, even mischievous, than Drysdale's brand, and in that sense more marginal. The *Kalogynomia* of 1821, by 'Thomas Bell', takes a libertarian line in its last section, but evidently as a mask for simple eroticisim (and the pornographic tradition has of course often espoused such an ethic, whether cynically or not). There is some show of intellectual gravity, but no more, in the anthropological comparisons with which James Lawrence in *The Empire of the Nairs* (1811) and Richard Monckton Milnes in the preface to *Palm Leaves* (1844) bolster their libertarian arguments. Henry Thomas Kitchener's *Letters on Marriage* (1812)—which I have discussed in Chapter 1—and then *Woman Physiologically Considered* (1839) by Alexander Walker, who like Drysdale was an Edinburgh medical man, are the two works of advanced sexual views which come closest to the *Elements* in tone.

Drysdale's enemies would have found his 'earnest' and 'philan-

[37] Nichols 1872: 301; 'Agate' 1885: 30, 53; *Weekly Dispatch* (22 Jan. 1860), 1; Gillespie 1869–72: 330.

thropic' spirit disconcertingly attendant on some of his most extreme views. The 'law of exercise' for the sexual organs, in both men and women, is a cardinal theme of his book: these organs, by analogy with others in the body, must be exercised by intercourse at least twice a week to be maintained in health. And the responsibility thus created, according to Drysdale, is not only intensely moral, but includes a duty to one's fellows, namely, to give them sexual gratification if they need it and you can offer it:

When once the commands of Nature are felt as they should be . . . *conscience will give youth no peace* till they be obeyed, and all obstacles to an honourable and disinterested outlet for the sexual desires surmounted. Every individual should make it his conscientious aim, that he or she should have sufficiency of love to satisfy the sexual demands of his nature, and that others around him should have the same.[38]

Such quotations may even give the impression that Drysdale's sexual code is just Victorian moralism with an interesting new twist, but his commitment to sexuality and sensuality, for their own sake, is quite genuine, if consistently serious-minded.

The emphasis in these remarks on young people's sexuality and its imperatives, for example, is characteristic: the *Elements* is very novel (though again in a sense Blakean), and prophetic, in its celebration of the era of late adolescence/young adulthood and its erotic rewards. Also, Drysdale agreed with many contemporaries that the human libido is modifiable, but almost uniquely he foresaw it becoming *stronger* as civilization progressed (and he held that in both sexes the more vigorous an individual's appetite for, and pleasure from, intercourse the greater his or her health). Drysdale is in fact extremely penetrating about the nineteenth century's valuation of the mind above the body, which he calls the 'spiritualism' of the age. Except for the important difference that he attributes this 'spiritualism' to Christianity, rather than to progressive rationalism, his account is like that offered in the present study; I would have adopted his term, rather than 'anti-sensualism', if it had not acquired misleading connotations later in the century.

The *Elements* opens with a wonderful clarion call of a sentence, which deserves a place in any dictionary of quotations: 'There is nothing from which mankind in the present day suffers more, than from the want of reverence for the human body.' Thus Drysdale

38 G. Drysdale 1868: 83–4, 366, 376.

broaches his first subject, 'Physical Religion', which the original title of 1855 (and the subtitle thereafter) indicates to be one of three 'religions' treated—with 'sexual' and 'natural' to follow. George Jacob Holyoake found that this terminology carried the Drysdalean seriousness too far; though a secularist, he objected to the sullying of religion with sex. (It is characteristic of Holyoake, however, that he has things both ways, and goes on to complain, inconsistently, that Drysdale's language might give 'physiological truths' an unfortunate 'reactionary' colouring.) The language of reverence is used very deliberately by Drysdale, and sometimes to astounding effect. Reverence for the body is not a narrowly hedonistic principle, but rather expresses the notion of an extreme humanist respect for every phase of our physical life: the ideal is that of 'a physiologically perfect life, perfect in every stage, perfect in its natural termination . . . The grand aim of the natural and only beautiful death should be kept steadily before the eyes of all throughout life.'[39]

Drysdale believed that all dead bodies should be reverently dissected. The function of medicine is nevertheless on the face of it called into question by the doctrine that our goal should be 'to live and die naturally, and to help others to do the same'. There are remarks on medicine in the *Elements* which are highly impressive to the modern reader for their emphasis on prevention, and on the doctor's duty of respect for the inner experience of patients, but perhaps in a full exposition of Drysdale's argument, which there is no room for here, he would have to be convicted of inconsistency in this area. There is a related, larger difficulty concerning the treatment of 'nature' and the 'natural' in the *Elements*. To put it no more strongly, the 'natural' is not for Drysdale benevolent, or even conducive to happiness, despite what one might gather from the above remarks, and similar ones about nature the 'all-perfect'. There is a sentence from the penultimate paragraph of the *Elements* as memorable as the one the book opens with: 'It is against Nature, not against each other, that we should contend; it is she, who is as omnipotent in the weapons of death, as of life—who has been our great destroyer.' But if Drysdale's notion of reverence cannot bear close logical scrutiny, piecemeal it produces some quite remarkable moments. The most impressive must be the calm and kind discussion of sodomy and lesbianism (in themselves virtually unmentionable topics in any kind of

---

[39] G. Drysdale 1868: 7, 36–7, 172–3, 328; see p. 171 above.

text at this period). Drysdale goes beyond even Bentham's advanced position on sexual deviance—namely, that society's treatment of it is absurdly 'out of proportion'—to a genuine tolerance, and application of the 'loving and reverential spirit' of his humanism.[40]

The earnestness of the *Elements of Social Science* is at once its strength and its weakness, for the intellectual scaffolding of his libertarian vision became, over time, too important for Drysdale. Together with certain associated interests this side of his system exclusively occupied his attention as a writer, in fact, once the first edition of the *Elements* had been published. The book's practical discussion of sexual morality and behaviour was never actively revisited, being reproduced unchanged and virtually unsupplemented in successive editions of the *Elements* for fifty years, and by the end of its life becoming very outdated.

The *Elements* was originally published anonymously with the title *Physical, Sexual and Natural Religion*. In later editions the date of this first version was always reputed to be 1854, but as no copies with this date on the title-page are known it is probable that those dated 1855 are actually the earliest, having been issued either very late in 1854 or at the beginning of 1855.[41] The '1854' preface was reprinted in later editions, and here Drysdale explained that he had published anonymously to avoid giving pain to a relation (his brother Charles later revealed that this relation was their mother, Lady Drysdale).[42] In 1855 the author is called 'a graduate of medicine'; in the tenth edition, of 1872, this is changed to 'a doctor of medicine' (Drysdale's name did not appear on the title-page until after his death in 1904).

Even in 1855 this was a long book—running to well over 400 pages—and it contained a great weight of medical and demographic material: about 150 pages on male and female sexual diseases (which are with great consistency ascribed to sexual abstinence), and some 60 pages of raw Malthusian theory in the form of a digest of the *Essay on Population*. By contrast, technical advice on birth control takes up no more than 3 pages. Within about a year Drysdale had created the new title, *The Elements of Social Science*. Holyoake, as I have mentioned, claimed responsibility for the change, but Drysdale explained that it was designed to indicate his admiration for the new science of society, by which he understood the whole range of

---

[40] Ibid. 7, 27, 130, 154, 181, 247–9, 288–9, 592.
[41] Himes 1928; Benn 1992: 282.   [42] C. Drysdale 1905.

allegedly empirical analysis of the subject from Comtean sociology through to the modern economic theory of Ricardo and Mill.[43]

He offered this defence in a short-lived periodical which he issued in 1856–7, significantly entitled *The Political Economist and Journal of Social Science* (though somewhat paradoxically its editor is referred to as 'The author of "Physical, Sexual and Natural Religion"'). This publication sets the tone for all Drysdale's post-1855 utterances on sexuality, with the emphasis heavily on the Malthusian underpinnings of contraceptive policy. In 1859 the *Elements* was enlarged even further with a 140-page section on 'Social Science', mainly culled from the *Political Economist*. The book had become preponderantly a rather heavy-going popularization of certain themes in political economy.

Drysdale was aware that there was a tension between his ambitious notion of the appropriate presentation of the argument for birth control, and the need to reach the millions of men and women oppressed by the effects of childbearing and sexual abstinence. He was actually urged by a correspondent in the *Political Economist* to produce a 6*d.* pamphlet shorn of all discussion of sexual disease and political economy, and he replied that he had considered such a solution, but had decided on the periodical format instead. The *Political Economist* retailed at 1*d.* for each monthly or two-monthly part of eight pages, which must have represented a considerable subsidy. It is possible that a penny pamphlet issued by Truelove in the late 1850s, *Poverty: Its Cause and Cure* by 'M.G.H.', is more than the work of an admirer of the *Elements*, and was to some extent drafted or devised by Drysdale himself. Norman Himes allows this possibility: Joseph Barker took *Poverty* to be an 'abstract' of the *Elements*. Truelove advertised it in the mid-1860s as 'highly recommended for extensive distribution by the Author of "The Elements of Social Science"' ('M.G.H.' had urged his readers to hand the pamphlet on to others or leave it lying in public places), and it may be vaguely alluded to in a review of the *Elements* in 1860 which speaks of how 'the substance of this volume has been in tracts circulated by the million among the operatives in the manufacturing and northern counties'.[44]

But the pamphlet would have been an isolated venture by Drysdale; certainly his preferred device for reconciling the needs of

[43] *Political Economist* (1856–7), 51–2.

[44] *Political Economist* (1856–7), 50–1; Himes 1936: 232; *National Reformer* (27 June 1861), 1; 'M.G.H.' 1870: 16; *Weekly Dispatch* (22 Jan. 1860), 1 (see also Mackay 1888: 28).

his argument and the needs of his audience was cheap or otherwise sponsored publication of larger scale texts. 'All his works were published at the lowest price possible', his brother recalled. The *Elements* was priced at 2*s.* and later at 2*s.* 6*d.*, a sum 'within the reach of most working men'; Holyoake, when he was in a mood to approve of the book, welcomed the way it was affordable by the 'poor'. It was, as another commentator observed, thus 'very cheap, but elaborate'. A scurrilous biography of Bradlaugh alleged that Drysdale directly remunerated those individuals who put their efforts behind the *Elements*, such as the publisher Edward Truelove and Bradlaugh himself. Joseph Barker said he had been 'offered a kind of bribe' to review it favourably. The numerous foreign translations of the book, into at least eleven languages, were apparently paid for by the author.[45]

The sales of the English language editions of the *Elements* followed the basic pattern of Owen's *Moral Physiology* and Knowlton's *Fruits of Philosophy*. The book sold about 750 copies annually for the first twelve years of its life, a level very like that achieved by *Moral Physiology*. Its circulation then accelerated to reach, by the mid-1870s, an average of 1,500 copies per year. This is actually somewhat better than the sales of *Fruits of Philosophy*, and the figures tend to support Bradlaughite claims that attacks on the *Elements* boosted its popularity.[46] The really big increase in sales, as with Owen and Knowlton, occurred at the time of the Bradlaugh-Besant trial. It is interesting, however, that Truelove was already putting out larger quantities of the *Elements* slightly before the trial: he issued two editions totalling 7,000 copies in 1876. The change in scale may have had something to do with the fact that *Fruits of Philosophy* was included as an appendix in some of Truelove's printings of the *Elements* at this date. This has not been noticed by historians of the Bradlaugh-Besant episode, and is rather intriguing: the legal vicissitudes of the more familiar Watts edition of *Fruits of Philosophy* were only just starting in December 1876.

For a dozen years after the trial the *Elements* sold about 3,500 copies a year. There was still a small and steady sale at the time of Drysdale's death in 1904, and for a few years thereafter, but not enough to justify resetting the text, although the publisher (now

[45] C. Drysdale 1905; Bradlaugh 1861: 10; *Reasoner*, 12 (25 Mar. 1855), 198; *Poverty* 1877: 25; Mackay 1888: 231–2; *National Reformer*, 8 (1866), 28–9.
[46] Field 1931: 219; *Bury Times* (29 Oct. 1870), 3.

George Standring) recognized that 'stereotypes which have seen fifty years' service' were producing an ugly result.[47] In its whole lifetime sales of the *Elements* fell somewhat short of 100,000 copies.

Late editions of the *Elements* were issued with three essays by the author bound in at the back: one was about statutory restraint of family size, and the other two discuss the prevention of war and infectious disease. Because the *Elements* was otherwise scarcely revised it became in due course an anachronism. There was never any reference, for example, to the availability of divorce from 1857, and the medical sections, in particular, became obsolete (though there was some attempt to update the discussion of syphilis). However the indications are that the author would not have altered the text of his book very much even if he had had the opportunity, and perhaps least of all in the sections concerning sexual function and relations. It is said that Drysdale monitored the foreign translations of the *Elements* very carefully, even to the point of teaching himself Russian. This activity, and the brief notes he furnished for the Bradlaugh-Besant reissue of *Fruits of Philosophy*, would appear to be his only return as a writer to the subject of sexuality in its physical aspect.[48]

One may deduce from his writings as 'G.R.' in the *National Reformer* in the 1860s that Drysdale continued to believe in such characteristic *Elements* doctrines as the 'law of exercise', the 'spiritual castration of our times', the pathogenic nature of celibacy, and reverence for the body. On these occasions the Drysdale of the *Elements* suddenly flashes out. But it is also true that the generally guarded rhetoric of 'G.R.' involves him, as views are exchanged in the pages of the *National Reformer*, in alignments that were alien to the spirit of the *Elements*. What did Drysdale feel when he was congratulated by a correspondent for promulgating the message that 'it is in not being master of his animal instincts that man is morally, as well as intellectually and physically, degraded'? He sat back for a considerable time while another contributor, David Hardie, attacked Malthusianism as the creed of sexual abstinence. He did eventually interject that modern Malthusians were bent on 'the radical extinction of celibacy and unhealthy continence', but oddly allowed Hardie, who was an eloquent opponent, the last word in this debate.[49]

---

[47] Fryer 1965: 314; C. Drysdale 1905: 'Publisher's Note'; Glass 1940: 40.

[48] C. Drysdale 1905.

[49] *National Reformer*, 1 (1861), 57, 5 (1864), 388–9, 6 (1865), 498, 752–3, 7 (1866), 359, 8 (1866), 162, 391, 9 (1867), 332, 390, 10 (1867), 31–2, 181, 363–4, 378–9, 11 (1868), 37–8, 67.

The surprisingly eirenic posture of 'G.R.' in these exchanges can partly be explained in terms of a general strategy adopted by Drysdale at this date. He was seeking in his journalism to forge a consensus on behalf of contraception, or the appearance of a consensus, by emphasizing a lowest common denominator of agreed views, while turning a blind eye to awkward fundamental divergences: even Matthew Arnold is discovered to be a neo-Malthusian. The strategy was an obvious descendant of Francis Place's manoeuvrings of forty years before, and in fact constitutes the essential position of all those campaigners for birth control who called themselves 'neo-Malthusians'. With George Drysdale, at least, it was more than a ploy. His *Political Economist*, for example, was partly a cheap, compact way of spreading the neo-Malthusian message, but also a forum in which its editor called for 'a *public demonstration*' on behalf of fertility control by a 'united' grouping of the working classes, 'democratic writers', and the 'liberal party among the economists'.[50]

This vision of a great neo-Malthusian alliance, made confident by a membership both broad and illustrious, came nearer to realization with the founding of the first Malthusian League, which occurred not with the setting-up of the familiar organization of that name in the wake of the Knowlton prosecution, but under Charles Bradlaugh's auspices in 1861. The fitful history of this eventually unsuccessful project can be traced in the pages of the *National Reformer*. Publicly it was Bradlaugh's venture, in the sense that its activities seem mainly to have consisted of lectures and pamphlets by Bradlaugh issued under the rubric of 'Malthusian League', backed up by some campaigning from a few provincial loyalists. But 'G.R.' was also a visible enthusiast, urging the need for 'an immense deal of public discussion' and exhorting people to join the League: 'it is numbers and an organized party of supporters that the doctrines need, to bring them publicly forward and prevent the press from ignoring a subject of such vital importance'.

This makes the thinking behind the League sound very like Drysdale's earlier plan for a birth-control union, and I believe that Bradlaugh was guided by the latter in his conception of the League (I shall touch shortly on the evidence for a special bond between the two men at this period). Norman Himes, who had some contact with the next generation of the Drysdale family in his pioneering research

---

[50] *National Reformer*, 12 (1868), 193–4; *Political Economist* (1856–7), 88, 119.

on the history of birth control, states bluntly that George 'founded' the first Malthusian League (and that 'the flower of his youthful enthusiasm withered from the cold' of its poor reception).[51]

The celebrity whose supposed support for birth control was most often cited by Drysdale was John Stuart Mill. Mill's remarks in *Political Economy* on the duty of the married not to have very large families (the means for this restriction not being specified) were standard pieces of ammunition with all neo-Malthusian propagandists, but in Drysdale's case loyalty to Mill went to extraordinary lengths, so much so that 'G.R.', at least, has more the appearance of a man supporting birth control because he believes in John Stuart Mill than vice versa. I have already noted the drift in Drysdale's activity in the mid-1850s towards an admiring emphasis on political economy. As 'G.R.' he filled pages of the *National Reformer* not only with expositions of Mill the economist but also with paraphrases of Mill's *Logic*, *Utilitarianism*, and *On Liberty*, or at least discussions of epistemology and ethics which are immensely dependent on these works.

Charles Drysdale remembered his brother making 'copious notes' on the classic political economists while he was still a medical student. Before this, according to the *Elements*, George was already haunted by the problems of sexuality: 'long before I read the works of Mr Malthus and Mr Mill, my mind was absorbed in the evils I saw and read of, from sexual abstinence, and other sexual difficulties and diseases'. At this period there was a luminary in George Drysdale's heaven at least as powerful as Malthus and Mill were later to be, namely, the French physician Lallemand. The latter receives the most unstinting praise in the *Elements* as the Newton of sexual medicine, sometimes in an intimate manner that suggests personal discipleship. However he came to be acquainted with Lallemand's work, Drysdale adhered to the latter's spermatorrhoeaic theory more closely than did any other English medical author.[52]

In short, George Drysdale gives the impression of being a man who formed almost fanatical intellectual allegiances. Annoyingly, there is little that can be done in the way of independently confirming this, or any other impression concerning this intriguing figure, because virtually no information survives about his life and character. This blank record is mainly the effect of Drysdale's own extreme reluctance for any kind of public exposure, of which the

[51] *National Reformer*, 8 (1866), 91, 9 (1867), 406; Himes 1936: 238.
[52] C. Drysdale 1905; G. Drysdale 1868: 96, 352.

anonymous publication of the *Elements*, to spare his mother's feelings, was just the first example. He adopted the pseudonym 'G.R.', or occasionally 'G.L.R.', for everything he published thereafter,[53] never took a formal role in the organization or leadership of neo-Malthusian activities, and certainly never spoke in public. If he attended meetings of the Dialectical Society, which was an important forum for advanced ideas on sexuality, he appears never to have contributed to the discussion (in all such contexts I take the bare 'Dr Drysdale' to refer to the very visible and active Charles). Drysdale never added to his Edinburgh MD and never published in the medical literature. Indeed, if he remained faithful to certain precepts about medicine expressed in the *Elements*—to the effect that doctors must go beyond 'the plain, matter-of-fact, physical treatment of disease' into prevention and psychology—it is hard to imagine that he would have worked as a mid-Victorian practitioner.[54]

The lack of easily exploited evidence has led to an almost complete silence from modern historians on a man who was by any measure an important Victorian. Only very recently has the Australian scholar Miriam Benn published the results of many years' painstaking efforts to get behind the blank record of Drysdale's life. She has made some brilliant discoveries by dint of the most grinding kind of search for facts that not even George Drysdale could be secretive about: such as his address and who was in his house on census night. Although she judges that he was 'not a recluse in the strict sense' he was obviously quite extraordinarily retiring. There is not a single reference to him in a social capacity in the published memoirs, diaries, and letters of the period. One portrait seems to survive but looks as if it dates from very early in his life (from Drysdale's beardless adolescence, in fact, if he was true to his 'physical religion' principle that men should not shave).[55]

Over the years various addresses are recorded for Drysdale in Churchill's *Medical Directory*. These are almost always the same central London addresses as those listed for his brother Charles, but it is now known that by the time of the 1871 census George was living in Penge (in the house where he was to die in 1904), and that by

---

[53] For a 'G.L.R.' item which seems clearly to be Drysdale's see *National Reformer*, 1 (1860), 31; the deduction in Ledbetter 1976: 11, 21–2, that he wrote as 'Q' in the *Malthusian* is based on a misreading of an announcement in that periodical.

[54] Nethercot 1961: 144; G. Drysdale 1868: 130.

[55] Benn 1992: 25; G. Drysdale 1868: 48–9.

1881 Charles was a near neighbour. It is a mystery where George Drysdale lived in the decade and a half after the publication of the *Elements*, while he was writing regularly for the *National Reformer*. He abruptly ceased to be a contributor in 1870, and there is a message for 'G.R.' in the *National Reformer* of 1875 which indicates that the editorship did not have a reliable address for this person.[56]

In view of the doctrines promoted in the *Elements* one is bound to be curious about George Drysdale's own sex-life. One could scarcely hope that anything could be recovered about this aspect of so elusive a career, and it is one of the remarkable achievements of Miriam Benn's research that she has shown that Drysdale cohabited with a woman of humble origins at least from his late forties. Letitia Radley, born like Drysdale himself in 1824, was the daughter of an agricultural worker from Suffolk. She had been briefly married to a local man, but widowed in 1852. It is not known when she met Drysdale, or when their common-law marriage started.

This discovery virtually disproves a favourite speculation of mine, that George Drysdale was sexually inactive, despite everything the *Elements* has to say about the importance of genital 'exercise' (there is also an anecdote in the unpublished autobiography of a progressive young doctor, Florence Fenwick Miller, which has Drysdale falling nervously in love with her when he was about 50). My main reason for guessing that Drysdale was sexually passive was a notable aside in the *Elements* about the compensations of being 'born sexually imperfect': 'every one who is born defective, stands in an exceptional position, the sexually neuter among the rest; and has experiences and perhaps opportunities of insight into nature, that others cannot have'. This reads strikingly like a description of the book's own achievement.[57]

Miriam Benn prefers to regard as autobiographical another arresting passage in the *Elements*, in which the case history is given of one of Lallemand's patients—driven to a breakdown by fears over masturbation, and cured by the French physician's prescription of regular sexual intercourse. This story is certainly given a striking weight in the text, and there is firm evidence of illness and breakdown interrupting George Drysdale's medical studies from 1843–6. He seems in fact to have deliberately dropped out of sight, for two years, while on holiday in Hungary, his family supposing him dead in a drowning

---

[56] Benn 1992: 149–51; *National Reformer*, 25 (1875), 345.
[57] Benn 1992: 151–4; G. Drysdale 1868: 190.

accident. It is less easy to make a case for Drysdale's having consulted Lallemand in Paris, however, as the 'case history' would require.[58]

'As much mystery is attached to the authorship', it was said of the *Elements of Social Science* in 1888, 'as, a generation and a half ago, was attached to the "Vestiges of Creation"'. How many people knew or guessed that the reclusive brother of Charles Drysdale was a controversial campaigner for sexual liberationism and birth control? How many even knew or guessed that 'G.R.' and the author of the *Elements* were one and the same? The answer seems to be that the second of these identities was often assumed (and could in fact have been deduced), while the more interesting truth about the flesh-and-blood reality behind both pseudonyms remained a well-guarded secret.

Despite the important differences between the *Elements* and contributions by 'G.R.' to the *National Reformer*, ordinary readers in 1863 could perceive them to have very similar doctrines, and entertain the thought that they were from a single hand. Truelove in his advertisements for Drysdale's various works gave the game away the following year when he announced *The Land Question* by 'G.R.' as a work by the author of the *Elements*, and the identity of the two was being taken for granted by 1872. It was made official by Bradlaugh and Besant in the 'Publishers' Preface' to *Fruits of Philosophy*.[59]

However, throughout these years, and down to George Drysdale's death in 1904, 'only a few people were privileged to know that he was the author' of the *Elements*. This group would have included individuals close to George Drysdale's projects, such as his two publishers Edward Truelove and George Standring and, I think, his editor Charles Bradlaugh, but also probably ideological allies such as the author of this remark, who was Paul Robin, the French neo-Malthusian leader. Annie Besant claimed in court in 1889 that she had never learnt who wrote the *Elements*, which strikes me as a perjury, but Bradlaugh's daughter Hypatia may not have known of Drysdale's activities as a writer even if she knew him as a family friend: he wrote three letters to her at the time of Bradlaugh's death signed 'G.R.'.[60]

[58] Benn 1992: 46–52.

[59] Mackay 1888: 230; *National Reformer* (31 Jan. 1863), 7, (25 Apr. 1863), 7, 20 (1872), 278–9; Chandrasekhar 1981: 90–1.

[60] *Malthusian*, 32 (1908), 9–13; *Daily Telegraph* (14 Nov. 1889), 5; Benn 1992: 25.

No one among the uninitiated seems to have penetrated Drysdale's disguise. The only hint to the contrary I have come across is Holyoake's writing in 1862 of 'the gentlemanly recluse who is the reputed author' of the *Elements*, but this phrase has a Russellian ambiguity. Holyoake could just mean that the author was reputed to be a gentlemanly recluse, not that a particular recluse was reputed to be the author. Crucial in this camouflaging operation was George's brother Charles, who was 'in complete agreement with his brother' ideologically, and shouldered all the public side of their work. Since the trail from the *Elements* easily led to the Drysdale doorstep Charles's reticence was a crucial protection for his brother, and allowed the curious to conclude that they had solved the mystery. Friend and foe alike, sometimes unhesitatingly, assumed that the 'graduate of medicine' and 'doctor of medicine' who had prescribed bi-weekly sexual intercourse for all in the *Elements of Social Science* was Dr Charles Drysdale MD, MRCP, FRCS.[61]

There are striking paradoxes in the behaviour of the Drysdale brothers, paradoxes of exposure and concealment, of boldness and inhibition. The author of the most comprehensively subversive treatise on sexuality in the English nineteenth century is one of the period's great recluses, or quasi-recluses. His brother was prepared to accept on his behalf the stigma of authorship of the notorious *Elements*, and yet, as I have described in the first part of this study, showed some reticence and even evasiveness on the details of birth control (and the Malthusian League, under Charles Drysdale's leadership, kept these matters at arm's length too). On the other hand, Charles had strong motives, it might be thought, for shunning personal attention as the supposed author of the *Elements*. It is another of Miriam Benn's discoveries that Charles Drysdale and Alice Vickery—who have always been known to be the parents of the birth-control campaigner Charles Vickery Drysdale—not only never married, but bore two children (Charles Vickery and the hitherto unknown George Vickery) while purporting to be living at separate addresses. They eventually let it be understood that they were married by starting to cohabit in 1895.

Miriam Benn has even doubted whether George Drysdale's determined anonymity as the author of the *Elements* really had anything to do with sparing the feelings of his mother, by arguing that the latter

[61] *Secular World* 1 (1862), 1; *Malthusian*, 32 (1908), 9; Mackay 1888: 27, 231; 'Agate' 1885: 21; Poulson 1892: 8; *Poverty* 1877: 8.

showed herself to be thick-skinned on sexual questions. Lady Drysdale was the mother-in-law of Edward Lane, correspondent in the Lane-Robinson divorce suit which I have described in Chapter 1. She gave evidence on Lane's behalf, an action which showed some capacity to expose herself to ugly sexual possibilities, but I do not think her endorsement of Lane's account of the affair indicates that she was 'realistic about sexuality' (in Miriam Benn's words), and certainly not that she was sympathetic to her son George's picture of our sexual economies. My impression is that Lane did have intercourse with Mrs Robinson, and that the allegations about the latter's 'uterine' delusions (which Lady Drysdale supported) were dishonest. Miriam Benn herself concedes that, in detail, these allegations were inconsistent with the principles of the *Elements*—since Mrs Robinson's mental disturbance was attributed partly to her contraceptive efforts.[62]

I believe that the leading doctrines of George Drysdale's book might well have been profoundly alien to his mother's beliefs about sex. To say this, of course, is to leave Charles Drysdale's tacit acceptance of many insinuations that *he* wrote the *Elements* no less puzzling. With both brothers an anti-sensual code, external to them but which they partly internalized, was perhaps having an effect. Whatever the case, there was actually, as I have argued at the beginning of this section, a new climate in the 1860s. The *Elements of Social Science* was out in the arena with its stark challenge of 'reverence for the human body'. This challenge evoked recognition and assent in many readers. The man who had uttered it was curiously absent and phantasmal, but there were others prepared to promote his vision.

It is not hard to find allegations that from about the time of the publication of the *Elements* there were many English men and women consciously committed, in some measure, to a libertarian sexual code. These allegations tend to be hostile in their attitude, and therefore probably exaggerate the phenomenon, but they are unlikely to be pure fantasy. It is certainly convincing, as *Tait's* reported in 1858, that premarital intercourse had its articulated creed, however furtive, and was 'inscribed with a sly pass-word, and accepted in whispers as a kind of right-wrong thing'. A few years later John Morley linked the 'nominal celibacy' of young men living in concubi-

[62] Benn 1992: 26–30, 147–51.

nage to the activity of 'writers' who had 'come boldly forward to
impugn' marriage, which gives some plausibility to the contemporary
claim that girls were being seduced with 'libertine language . . .
about love being "supreme", marriage "a humbug", desire "active
love", and so forth'. Justin McCarthy also judged that the spread of
'immorality' as a 'general institution' was a principled matter, the
result of modern self-scrutiny: 'marriage might almost seem to be . . .
just now upon its trial'. Diana Mulock deplored the indications that
respectable female opinion was being infected by 'what licentious
materialists are now-a-days unblushingly asserting, that chastity is
not indispensable in our sex'.[63]

Some of these middle-class writers fairly evidently have the
Elements of Social Science in mind (William Acton said as early as
1857 that it was 'the mouthpiece of a large class of young men'),[64] and
it would be interesting to know how much they knew of the book. It
was never generally available, and of the ten or so English periodicals
or newspapers which carried reviews none would have come readily
before the notice of middle-class readers. There is no mystery, however, about the currency of the *Elements* with the audience it was
more directly aimed at, that is, Drysdale's union of the working
classes, 'democratic writers', and 'liberal' economists. In the mainstream radical press the book received just a handful of notices, over a
long period, and of these only a review in the Chartist *People's Paper*,
inexplicably enthusiastic, avoided reflex radical dislike for Malthusian
ideas.[65] It was almost entirely through secularist agencies, and mainly
as these were courageously deployed by Charles Bradlaugh, that the
Elements of Social Science was kept before its potential readership for
several decades, and achieved its growing influence.

Bradlaugh appears to have been a neo-Malthusian before the publication of the *Elements*, perhaps from as early as 1852. When he
started to edit the *National Reformer* in 1860 he wrote a series of articles, most of them issued also as pamphlets, urging birth control as a
means of political and social advance for the working class: the first
and most important was *Jesus, Shelley and Malthus* of 1861. While
the channel offered by the magazine was one reason for this burst of

---

[63] 'The Ethics of Early and Frugal Marriages', *Taits*, 25 (1858), 97–100; Morley 1862;
Social Science 1861: 67; McCarthy 1864; Mulock 1858: 225.

[64] Acton 1857a: 13–14.

[65] *Weekly Dispatch* (22 Jan. 1860), 1, (29 Jan. 1860), 1, (5 Feb. 1860), 1; *People's Paper* (24
Feb. 1855), 5; *Examiner* (4 Jan. 1873), 14.

activity from Bradlaugh Drysdale's ideas were probably a necessary catalyst. Extensive excerpts from the *Elements* had been printed in the *Investigator*, which Bradlaugh edited just before launching the *National Reformer*; as soon as the latter was under way he inserted an unequivocal recommendation of Drysdale's book, and a year later endorsed it more conspicuously in *Jesus, Shelley and Malthus*. I have already suggested that the plan for a Malthusian League, of this date, came from Drysdale. There is a striking warmth in the tribute which Bradlaugh offered 'G.R.' in March 1862; it explicitly goes beyond an editor's gratitude to a valued contributor, to record a personal debt: 'to the writer of those articles on Political Economy which have provoked the ire of our unco guid opponents, our acknowledgements are specially due. We are personally deeply indebted to G.R., for that he has opened to us a wide field of possible usefulness from which no vulgar insinuations shall turn us away.'[66]

To speak of 'G.R.' as arousing the wrath of the 'unco guid' really only makes sense if Bradlaugh has at the back of his mind Drysdale in his other manifestation, as the author of the *Elements*—a book which had, in particular, come under tremendous assault the previous year from both Bradlaugh's co-editors on the *National Reformer*, Joseph Barker and Holyoake (there was no comparable response to the journalism of G.R.). Bradlaugh is in effect associating himself with the transgressive aspect of Drysdale's sexual thought—as he also does simply in continuing to endorse the *Elements* after 1861. Moreover in his many supportive remarks about the book right through the 1860s and 1870s, the nearest Bradlaugh comes to criticizing its moral views is to say that it contains 'some points of ethics on which I dissent' and that 'on the marriage question, the author must be read with extreme care, as the views he raises are so different from ordinary public opinion'. Given the extraordinarily liberal doctrines of the *Elements*, and the bitter hostility they aroused in some quarters, this is to praise with faint damns. The enemies of the book perceived Bradlaugh and his colleagues to be engaged in a vigorous campaign to promote its popularity throughout the country.[67]

[66] Bonner 1895: ii. 172 ('1862' here seems to a misprint for '1852', see also Chandrasekhar 1981: 50); *Investigator* (15 Mar. 1859), 199; *National Reformer*, 1 (1860–1), 7, (8 Mar. 1862), 5; Bradlaugh 1861: 4, 10.

[67] *National Reformer*, 9 (1867), 329, 11 (1868), 153, 12 (1868), 105, 13 (1869), 198–9, 15 (1870), 386–7; Gillespie 1869–72: 322–38; Bradlaugh 1861: 10; Barker 1869: 123; Mackay 1888: 28. Bradlaugh's remarks are misrepresented by Tribe 1971: 336 and by Benn 1992: 237 following Tribe.

It seems likely that in his personal sexual morality Bradlaugh was no practiser of the 'law of exercise'. His modern biographers, and Annie Besant's, have judged that the couple's relationship—though certainly erotic, to the point where Annie spoke of herself as 'engaged' to Bradlaugh—was platonic. Associates of Bradlaugh, such as G. W. Foote and Annie herself, depicted a man strongly committed to monogamous marriage as an institution, and thus holding views the 'exact contrary' of Drysdale's. But the fact is that Bradlaugh, even when hard-pressed in public debate on his movement's loyalty to the *Elements*, for many years would only express a limited degree of dissent from the book's moral position: denying that 'seduction was a physiological virtue', for example, but suggesting that such readings of the *Elements* were a distortion, and urging that it had been 'penned with a pure intention'. There was admittedly a shift in his tone concerning the *Elements* in the early 1880s, almost certainly as a defensive response to the way the sexual card had been played by those opposing his becoming MP for Northampton (where he had won the seat in 1880). Bradlaugh now distanced himself more emphatically from libertarian views when he addressed fellow secularists, and he added a footnote to the 1883 edition of *Jesus, Shelley, and Malthus* in which he spoke of 'some points of ethics' in the *Elements* 'from which I expressly and strongly dissent'.[68]

I shall suggest shortly that this did not amount to a defeat for the libertarian position, though it has a similar appearance to Robert Owen's climb-down of forty-five years earlier. In that episode there was very significant opposition within a progressive movement to any relaxation of sexual codes. Bradlaugh's antagonists in the 1880s continually rehashed stories of secularist disagreement with the leader's partisanship of the *Elements* in the 1850s and 1860s. Some of these attempts to depict a movement split from top to bottom on sexual questions were plainly wrong: as when Bradlaugh was represented as having hijacked the *Investigator* in 1858 to promote the *Elements*, although the book had actually been given a rather fuller endorsement by the magazine's previous editor, W. H. Johnson, than under Bradlaugh's control. Johnson was evidently acquainted with the whole text of the *Elements* (and not, as was later alleged, only with the 'Malthusian part' of the proofs); he appears to have reviewed it in the teeth of attempts by other secularists to deter him, and was then

---

[68] A. Taylor 1992: 91–3, 124–5; Tribe 1971: 180; *Bury Times* (29 Oct. 1870), 3; Foote 1895: 4; Besant 1881: 10–11; Bonner 1895: ii. 160–1.

abused for having done so (according to the later version he simply received chastening advice when the review appeared).

By the same token there obviously *were* secularists who strongly disapproved of any association of their movement with Drysdale's doctrines. Some of the fiercest attacks of the 1880s on Bradlaugh's supposed libertarianism come from free-thinkers such as G. C. Griffith-Jones and W. S. Ross. (By a typical twist, W. H. Johnson himself was one of the secularists who in 1888 hid their identities behind that of the hack Charles Mackay in a scurrilous life of Bradlaugh: Johnson did previously express regret for having praised the *Elements*.) I have discussed Holyoake's wavering hostility to the *Elements*, and mentioned Joseph Barker's attacks on it. It would obviously be very wrong to suggest that Drysdale's book, and the free-love doctrines it promoted, were other than a major source of friction within secularism. What can be said as to the balance of opinion in the controversy it fomented?[69]

The liberal Unitarian P. W. Perfitt, surveying the quarrels concerning the *Elements* in 1861, distinguished 'three parties' of secularists:

The first is composed of men who make unceasing war upon the entire circle of theological systems . . . they repudiate Inspiration, and its cognates, but uphold Morality with all its bonds and issues . . . Such men shrink from the book in question; to them there is pollution even in its touch . . . The opposite class to these is composed of low-browed, coarse, uneducated and sensual men . . . Their tastes are as low as their brows, and if they do not fight cocks or draw badgers, it is not that their tastes are improved, but that the law is too strong against them . . . Between these two classes there is yet a third. They are neither given to vice nor desirous of having it advocated, but they are so wedded to Secularism, that they slavishly tolerate movements which their consciences cannot approve; and they cannot pardon the man who publicly questions the propriety of any Secularist's conduct.[70]

In this wonderfully biased account there is no mistaking where Perfitt himself stands, and his whole analysis is something of a travesty. But the notion of two wings of secularism, one moralistically hostile to the *Elements* and the other a staunch supporter, with an intermediate group compromised into partial loyalty, is a useful starting-point, especially if the outline of the compromise party is allowed to be blurred at both its edges, and the possibility recognized

[69] Mackay 1888: 26–8, 231; *Investigator*, 4 (1857), 77–9; G. J. Holyoake ('Quasimodo') 1861: 2; *Barker's Review*, 1 (1861), 11, 205; Benn 1992: 19–23.

[70] *Pathfinder*, 6 (1861), 30.

that under sufficient pressure the compromisers could cease to be 'slavishly' libertarian.

One aspect of the *Elements* which tended to make compromisers out of even its most decided enemies was its elaborate commitment to birth control. Between 1830 and 1877, a modern historian has concluded, conservative neo-Malthusianism (that is, a policy of birth control practised within a more or less orthodox marriage-code) became 'a generally accepted principle of the secularist philosophy'. I have noted already how this development contributed to the uncertainty of Annie Besant's and Holyoake's hostility to the *Elements*, and how it potentially undermined Francis Newman's. Even Joseph Barker hinted that he could accept some kind of fertility-restriction other than Malthusian 'moral restraint', and that it was the hedonistic application of contraceptive principles in the *Elements* which had obliged him to reject the book wholesale. He also conceded that the celibacy and marital unhappiness highlighted by libertarians were indeed acute problems.

Barker was certainly the fiercest of the early opponents of the Drysdale–Bradlaugh alliance, but his secularism, indeed his whole metaphysical affiliation, was so bizarrely unstable that he cannot be thought of as representative. It was possible for Nonconformists to give considered approval to birth control (in 1878 a Birmingham congregation voted 23 to 7 in favour), but Barker in his religious phases was no ordinary Nonconformist. Before 1860 he had been a prominent Methodist preacher and anti-Owenite, and then—repenting of his attacks on socialism—leader of his own Unitarian-minded sect of 'Barkerites'. After the rift with Bradlaugh he became a Coleridgean Anglican, a Spiritualist, and a Methodist again.

Barker's anti-sensualism (rather like that of another ex-Owenite opponent of the *Elements*, T. L. Nichols) belongs more with the ambiguous anti-sensualism of religious sectarians: though the Barkerites had an austere code, Barker as a publisher (another of his manifestations) dabbled in sexual material. In 1863 he wrote a booklet on marriage which waxes warm on the 'ecstatic' and 'rapturous' pleasures of love. Barker was accused pseudonymously by Holyoake (though this perhaps says more about the strange contortions of secularist rancour) of having turned the *National Reformer* into 'the most disgusting paper which ever disgraced the literature of Freethinking' by printing a couple of items of sexual advice.[71]

[71] Micklewright 1962; *National Reformer*, 1 (1860–1), 363, (3 Aug.), 4, (17 Aug.), 5, 31 (1878), 1142; *Barker's Review*, 3 (1862), 107–10; G. J. Holyoake 1875: i. 327–8; Rushton

There were of course some uncompromised enemies of the *Elements of Social Science*, root and branch, from the start: men such as William Maccall and John Maughan (whose opposition to birth control contained more than a vestige of the old radical attitude), and W. S. Ross, who early in his career spoke out vehemently against Bradlaugh's sexual doctrines at a meeting when he saw Holyoake 'shilly-shallying', Charles Watts also evasive, and only G. W. Foote taking a firmly moral line.[72] Watts was the nervous publisher of the *Fruits of Philosophy*, who had been easily frightened by legal action out of further loyalty to the book, and the celebrated events of 1877 were in some sense a test of the strength of secularist libertarianism. Knowlton's book, as I have said, was transgressive in its underlying moral code, and in the Bradlaugh-Besant reissue it had notes expressly credited to the author of the *Elements of Social Science*. The contemporary evidence, and the verdict of later commentators, is that though the decision to champion Knowlton and birth control in the courts put a severe strain on secularism, and certainly produced some schismatics, Bradlaugh emerged at the head of an intact movement.[73]

On the other hand, it was first and foremost contraception which was on trial in 1877, and there *were* defectors. If allegiance to the *Elements of Social Science* had been put to a similar test there is no reason to doubt that free-thinkers would have repudiated the more libertarian work in rather larger numbers. I mentioned at the beginning of this discussion the case of the free-thought lecturer, Harriet Law, being forced into such a repudiation by a public challenge. The *Elements* has several recorded champions, but only a proportion of these expressly endorse the book's free-love doctrines[74]—and it would be unwise to suppose that the remainder went all the way with Drysdale on the duty of sexual promiscuity, in particular.[75] It is

---

1909: ch. 15; Lawson 1887: 76–82; Podmore 1906: 501–2; Budd 1977: 55; Royle 1974: 307; Barker 1839: *passim*, 1846: *passim*, 1858: *passim*, 1863: 8, 1869: *passim*; G. J. Holyoake ('Quasimodo') 1861: 5; Nichols 1872: 300–1, 1886: 163–7; Aspinwall 1982.

[72] Royle 1974: 277–8; Tribe 1971: 76; *National Reformer*, 1 (1860–1), 238, (1 Nov. 1862), 5, 10 (1867), 299; 'Agate' 1885: 5–6.

[73] Budd 1977: 56–60; Nethercot 1961: 131–2; W. S. Smith 1967: 49–52; *National Reformer*, 30 (1877), 644; Tribe 1971: 176–7, 184; F. B. Smith 1967; Micklewright 1962; McCabe 1908: ii. 85.

[74] 'Philanthropus; 1879: *passim*; Harte 1870: *passim*; Clarke 1962: 180; Gould 1923: 35; see also 'Gordius', *National Reformer*, 13 (1869), 247, 'Oxoniensis' 1883: pp. i–99.

[75] *Poverty* 1877: *passim*; A. Holyoake 1870: 'M. A., Cantab', *National Reformer*, 18 (1871), 53.

known that Edward Truelove, who took the book on when no other
publisher would, and loyally continued with it until his death, none
the less had reservations about it. The *Elements* is praised in a work
Truelove himself stood trial for having published in 1878, J. H.
Palmer's *Individual, Family and National Poverty*, but while the
author agrees with the Drysdalean law of genital exercise he does *not*
accept the need for premarital intercourse: for the male nocturnal
emissions are, according to Palmer, an adequate substitute.[76]

There was one celebrated instance of an individual's being
bounced out of his allegiance to Drysdale. The Dialectical Society,
about which regrettably little is known, was in principle a London-
based discussion group, active through the last third of the century,
with no axe to grind, dedicated only to the free exchange of opinion.
In practice it was a forum for ideas too advanced to be voiced anywhere else (except perhaps in the even more obscure Eleusis Club).
It is chiefly remembered today as an early haunt of George Bernard
Shaw's. Many of its members were women, who participated freely
in the debates.[77] There was no party line on sexuality at the
Dialectical Society, but the desirability of birth control was 'accepted
as a self-evident proposition by almost all the members', and there
were certainly those who believed in the physical necessity for intercourse, and in free sexual liaisons. There was, by contrast, also a
strong survival of the liberal rationalist belief in the decay of the
libido with the advance of mental culture.[78]

Lord Amberley was a Dialectician caught between the old progressive anti-sensualism and the new Drysdalean hedonism: in some
measure an ally of John Stuart Mill and his belief that the sexual
'propensity' was fast becoming 'completely under the control of reason', but also an admirer of Drysdale's *Elements*, at least for its candour if not for its 'materialistic notions'. He may or may not have
recommended the book during a Dialectical Society discussion
which he chaired on 1 July 1868, but any inclination he felt subse-

[76] C. Drysdale 1905; G. Drysdale 1868: p. v; Palmer 1878: 14, 17.

[77] For information on the Dialectical Society see its few surviving publications, and
Holyoake diaries (Bishopsgate Institute); Nethercot 1961: *passim*; W. S. Smith 1967: 9–10;
Davies 1874 i. 137–40, 176–207; Bax 1918: 227–8; Lea 1881: 1; G. B. Shaw 1949: 56; Tribe
1971: *passim*. The society's activities are noted in *National Reformer* and its debates on sex-
ual questions often reported in *British Medical Journal, Public Health, Medical Enquirer,
Medical Press and Circular*. Somewhat earlier, probably, there was a Dialectical Society in
Glasgow (see *Biograph and Review*, 2 (1879), 155–63).

[78] Coupland 1878: 14 and *passim*; *Medical Press and Circular*, 1870(1), 211–12; *Public
Health*, 1 (1868), 169–70; Davies 1874: i. 185, 190; Bax 1918: 227.

quently to endorse it, or to endorse even birth control (which he certainly had spoken up for at the meeting), was swept away by a tide of public indignation. In reality this emanated mainly from doctors who hated his calling on their profession to purvey contraceptive wisdom, and from Amberley's political opponents, who saw the opportunity to foil his bid to be Liberal member for South Devon in the election of 1868 (the son seeking to follow his illustrious father, Lord John Russell).

The effect was to elicit from Amberley the most vehement repudiations of the *Elements* and its doctrines, and denial that he had ever been other than disgusted by the book. He even denied what was well-attested: that he had called upon doctors to enter the forum on contraception. He was not the only man to be intimidated to this extent by the reaction to reports of the 1868 meeting: the Dialectical Society's secretary made out in the *Lancet*, with extreme inaccuracy, that all his members were opposed to birth control.[79]

This episode, especially if put together with Bradlaugh's very much slighter but still discernible shift of tone on the *Elements* in 1880, may look like a defeat for liberal sexual policies (moreover, given the apparent effectiveness of electoral opinion in both cases, a defeat at the hand of a rather broad segment of society). But the successful mobilizing of anti-sensual forces was accompanied by an equivalent mobilizing on the pro-sensual side. Most significantly, the old tradition of progressive anti-sensualism tended to be squeezed out of the picture. When Amberley and his like declare themselves enemies of liberationism their anti-sensualism is swallowed up in the conservative and religiose moralism which became so powerful in the last two decades of the century. The old working alliance between classic moralism and the progressive temper, which had given such unanimity to the sexual culture of England for decades, was thus broken.

The terms of the debate had changed. Men and women were faced with a starker set of alternatives for the conduct of their sexual lives: on premarital sexuality, on birth control, and simply on the general valuation of sexual pleasure. Lionel Beale, in 1887, comments very interestingly on this new division of codes. On one hand, he says, 'the standard of morality held by many is higher than it was two

---

[79] J. S. Mill 1972: 1692–4; Russell and Russell 1937: ii. 242–5; *Medical Press and Circular*, 1868(2), 84–5, 210–11; *Public Health*, 1 (1868), 208–11; *BMJ* 1868(2), 113; Besant 1881: 10; *Medical Times and Gazette*, 1868(2), 244; *Lancet*, 1869(2), p. 321.

generations ago' (and as represented by the social purity movement this moralism had gone too far for Beale's taste). However, over the same period a completely novel set of 'monstrous and untenable propositions' has been deliberately advanced by certain 'physiologists and philosophers', to the effect that 'man's sexual instincts . . . are not to be curbed, resisted, or disobeyed'. This campaign, according to Beale, had had genuinely erosive consequences for 'moral opinion and practice' in society: 'an immoral minority has succeeded in demoralizing large numbers'.[80]

It is significant that the man who sought to exploit the story of Bradlaugh's association with the *Elements of Social Science* most energetically in the 1880s was Henry Varley, pioneer social purity lecturer and pamphleteer. But even before Varley's interventions the social purity movement had served to catalyse the statement of libertarian views on sexuality by secularists, sometimes in terms that seem to be directly inherited from the *Elements*:

We acknowledge no distinction of higher and lower . . . The passions they [the Social Purity Alliance] would fain reduce to almost imperceptible strength, are as needful a part of human nature, and as much deserving of cultivation, as any impulse . . . the sexual part of our nature . . . is intimately connected with what they are pleased to call the higher parts . . . the strength of it is a measure of the strength of the whole nature.

Such writing suggests fascinatingly how Drysdale's *Elements* set the agenda for late nineteenth-century libertarianism. Another example is the neo-Malthusian J. H. Levy's moral defence of the prostitute: thirty-five years earlier Drysdale had urged his readers to feel 'deep gratitude' towards the prostitute for her beneficial sexual activity, and 'reverence' towards her 'as for all other human beings' for the 'wonder and mystery' of her nature.[81]

There is a similar story of antagonisms sharpened, and of positions becoming more entrenched, in the case of birth control. In the late 1860s there was for the first time an open discussion among doctors on this subject. Almost all of them were strongly opposed to any involvement by their profession in contraceptive practices, and as this medical antagonism developed in the wake of the Knowlton affair there entered—especially in the propaganda of C. H. F. Routh—a new shrillness and moralistic extremism on behalf of physi-

---

[80] Beale 1887: 47, 53, 57–9, 68, 97–9, 127–9, 137.
[81] *National Reformer*, 31 (1878), 887, 33 (1879), 275–6; G. Drysdale 1868: 270–1.

cal asceticism and self-discipline. An active campaign by both doctors and lay authorities was launched against the advocates of birth control, on the face of it with considerable success. The Bradlaugh-Besant prosecution failed on a technicality, but that of Edward Truelove, a year later, for publishing Robert Dale Owen and J. H. Palmer, succeeded, and the 70-year-old free-thinker served a term of imprisonment with hard labour. Similar actions were being brought right through the turn of the century. Most notably, the Yorkshire practitioner Henry Allbutt, author of *The Wife's Handbook* (1886), was stripped of his licentiates and fellowships, and struck off the medical register—while there were some prosecutions of individuals for selling his booklet.

It is clear to us now that all these were hollow triumphs. The spread of contraception was irresistible, and the early prosecutions backfired badly in the sense that the Knowlton trial received huge publicity, and Truelove's plight created much indignation. Allbutt's book went from strength to strength in the market-place, while he himself became prosperous and sought-after as a practitioner.[82] It is hardly less contestable that the victories of anti-sensualism in the area of sexual codes, at this date, were also illusory. Men and women faced with a choice between sexual hedonism and its doctrinaire opposite in due course chose the former. The Social Purity Alliance was a grotesque affair, a great bubble of extreme anti-sensualism whose only lasting achievement was to discredit sexual moralism. In fact, it is surely this episode of last-gasp, hypertrophied asceticism, occurring at the very end of Victoria's reign, which largely explains the very swift establishment of the 'Victorian' as a stigmatizing label in the first twenty years of the next century. The *Elements of Social Science* was the most influential libertarian statement of its age, and a major stimulus, directly and indirectly, to the social purity operation. In that sense George Drysdale may be said to have destroyed the Victorian, and created the 'Victorian'.[83]

[82] Routh 1879: 9–10 and *passim*; Peel 1963; D'Arcy 1977; Micklewright 1962; Ledbetter 1976: 131–41; Banks and Banks 1954; Nethercot 1961: 136; Himes 1936: 251–6.
[83] Cf. Cominos 1963.

Epilogue

WOMEN AND CHILDREN

As the public debate on sexuality polarized at the end of the nineteenth century, into virtually unprecedented extremes of libertarian and anti-libertarian policy, one radical faction became attached to
the latter of these extremes. For two decades or so there were close
links between the social purity movement and the agitation for
women's rights. This is an especially striking instance of a phenomenon which, in other manifestations, I have sought to illustrate
throughout this study, and it has been carefully explored by historians of feminism in recent years. These investigators are, indeed,
unique in having noticed and tackled the intimate connection
between progressive and anti-sensualist thought in their field: elsewhere this very distinctive nineteenth-century constellation of attitudes has hardly been acknowledged. In the case of feminist
anti-sensualism I can leave the story to several expert voices.

All these voices have been female, and more or less actively feminist. For these writers there has been a particularly strong induce-
ment to do what is natural for any historian, I think, who starts to
notice the progressive colouring of the 'Victorian', namely, to reflect
on the message it may have for our culture of a century later—a culture which is the undoubted beneficiary of nineteenth-century progressive thinking, but at the same time thoroughly opposed to the
period's ethic of sexual reticence and restraint. Most women historians writing in recent years about the pronounced anti-sensualism of

the women's movement in the years 1880–1920 have shown an unease about the phenomenon, but increasingly it has been analysed without condemnation, and even approvingly.

The first book to tackle the topic directly was Constance Rover's *Love, Morals and the Feminists*, which appeared just over twenty years ago. Rover was frankly alienated by the suffragists' emphasis on sexual purity, and their hostility to liberalized sexual codes and to birth control. She explained these features of their programme as a kind of ideological infection from their social background, these women being 'bourgeois minded' (more recently a male historian, Frank Mort, has offered a related argument: that feminist anti-sensualism at this period—especially as it manifested itself in efforts to root out prostitution—was driven by class interests). A variant of this account is that the feminist leadership became more 'conservative' on sexual questions as it achieved increasing power and standing towards the end of the century. (And it has been argued that this conservative colouration arose because feminists were not able to prevent their successful campaign against the double standard from being hijacked by groups with 'antithetical' interests).[1]

This is to explain away the anti-sensualism of these women as adventitious to their feminism: as the product of some other factor which they shared with each other, such as their social origins. Essentially the same logic, though invoking a different background, is to be found in Olive Banks's *Faces of Feminism*. Banks, in a classic study with her husband over a quarter of a century ago, first drew attention to one of the most striking aspects of this whole question: that birth control was undoubtedly gaining ground very rapidly as a component of marital sexuality in the last third of the century, but was consistently ignored or deplored by feminists of the day. In her more recent book Olive Banks has accounted for the austere sexual code of many suffragists as an effect of the evangelical religion which they were still swayed by, as 'an aspect of the religious tradition'.[2]

One case which is awkward for the claim that these Victorian and Edwardian feminists were anti-sensual because of some aspect of their background is the colourful example of the Pankhursts. Lady Emmeline Pankhurst, and then—of her two feminist daughters—Christabel, were both classic high-minded activists where sexual issues were concerned: opposed to birth control, censorious on non-

---

[1] Rover 1970: 31–5; Mort 1985; DuBois and Gordon 1984; Walkowitz 1982.
[2] O. Banks 1981: 63.

marital liaisons, enthusiasts for male chastity and continence, inclined to a low estimate of women's sexual drives. But Sylvia Pankhurst, Emmeline's other daughter and no less of a feminist, was in principle and practice opposed to all these attitudes. 'We are for free love', she wrote in her notebook: 'polyandry . . . is legitimate'. She never married but probably had at least two serious affairs, and bore an illegitimate child. She defended her conduct in the *News of the World*, but relations with her deeply disapproving mother and sister were broken for ever.[3]

It may be objected that to cite the Pankhursts is to make no more than a debating point. The fact that the women of one family could diverge so widely on the sexual content of female emancipation does not in itself rule out cultural or social background as the main source of the feminist anti-sensualism of the day. Not everyone conforms to their background. It could at least be argued that the deep rupture between the Pankhurst women proves the neutrality of the feminist creed on sexual issues—and indicates that differences on this score must be referred to other factors.

Another dramatic case may be brought in evidence. In 'The Girl of the Period' Eliza Lynn Linton wrote the century's most celebrated jeremiad about the evils of any departure, by women, from a strict sexual code, and she published it in the great organ of traditional moralism, the *Saturday Review*, in 1868. Linton had been a feminist, and in due course became a vehement opponent of women's rights. It is interesting that this shift did not move in step with Linton's growing anti-sensualism. If anything the connection worked in the opposite direction for, in true *Saturday Review* style, she hated typical feminist purity goals such as male continence and abolition of the Contagious Diseases Acts. But there are signs that Eliza Lynn Linton's feminism was always an insecure creed, and her remarkable affiliation to *Saturday Review* male chauvinist moralism may not be relevant to the proposition that feminism was neutral in its influence on sexual attitudes. Feminism proper could produce wildly divergent creeds, as with the Pankhursts, but the pattern of outcomes was, so to speak, bi-stable—very much in the manner I have already traced in the case of Owenism, for example. If Sylvia Pankhurst is the William Hodson of feminism Christabel is the Goodwyn Barmby.[4]

Some of the most recent historians of feminism have in fact been

[3] Romero 1987: 118, 168–71; D. Mitchell 1967: 69–72.
[4] Gay 1984: 210–2; Anderson 1987: 198–200, 1989; Helsinger *et al.* 1983: i. 103–25.

prepared to embrace the thought that the anti-sensualism of the 1880–1920 period—though in its 'Victorianism' it goes as far as, or even further than, anything from the rest of the nineteenth century—was intrinsic to the movement. The more cautious have interpreted this moral colouration as a mere strategy, as a 'subversive' adopting of doctrines of female moral superiority in order to wrongfoot the reactionary opinion which cherished such doctrines. Bolder accounts have taken this feminist anti-sensualism at face value, and spelt out, non-judgementally, the logic of their case against sex as feminists argued it. A central theme of this case was that the agenda of sexual 'liberation'—free liaisons, birth control, hedonistic sex—was self-servingly devised by men, and spurious in its apparent concern for women.[5] These claims—about which the first generation of late-Victorian feminists, at least, seems to have been unanimous—are disconcertingly impressive. They may have been called forth by a polemical purpose, but they are uttered by the first women in the period, and arguably the first in English history, to report on their experience of sex.

The most unabashed of the recent historians has certainly been the radical feminist Sheila Jeffreys. She makes no bones about late-Victorian feminism's involvement in the most coercively moralistic propaganda of the day ('feminist ideas and personnel played a vitally important part in the 1880s social purity movement and . . . shaped its directions and concerns'), and is not embarrassed by the company of such admittedly reactionary figures as Ellice Hopkins. The views of women like Elizabeth Wolstoneholme Elmy and Francis Swiney—who held that reproduction should be the only motive for physical sex between men and women—are sympathetically reported, and other historians chided for finding them a 'stumbling block'. Jeffreys entirely accepts the argument that 'free love' and its adjuncts were a stalking-horse for male gratification, and she has since carried the case against the 'baloney of liberation' into the twentieth century and down to the present day.[6]

Those who would seek to explain away feminist sexual moralism as a social or cultural accident, or even as a cunning strategy to disarm the enemy, have to explain away much more than this late-Victorian episode. Anti-sensualism is an almost inalienable feature of the femi-

---

[5] Levine 1987: 132–3, 1990: ch. 5; S. K. Kent 1987: 92–111; Shanley 1989: 184–67. See also Cott 1978.

[6] Jeffreys 1985: 9–44; 1990: 2 and *passim*.

nist tradition in the nineteenth century. In introducing its founding
text, *Vindication of the Rights of Woman*, Mary Wollstonecraft says
that she has 'produced many arguments' tending to show that
'chastity must more universally prevail', but this scarcely does justice
to the astonishing weight of anti-sensualist precept in the book that
follows. At every turn of the discussion women's sexuality—negatively regarded—is a central preoccupation. Chapter 7 of the
Vindication even opens with an invocation to 'Modesty! sacred offspring of sensibility and reason!' in frankly Miltonic style.

Wollstonecraft wishes that chastity 'more universally prevail' to the
point of breaching the walls of matrimony itself, so that married couples are urged to be as continent as possible: 'a master and mistress
of a family ought not to continue to love each other with passion';
'when even two virtuous young people marry, it would perhaps be
happy if some circumstances checked their passion'. This anticipates
the most austere doctrines of Elizabeth Elmy almost a century later,
and Wollstonecraft's version of feminist anti-sensualism actually outdoes the social purity feminists in the sense that sexual continence is
argued to be *more* incumbent on women than on men (whereas the
social purity activists wanted a *quid pro quo* of sexual restraint from
men). 'Whether she be loved or neglected', says Wollstonecraft of the
ideal wife, 'her first wish should be to make herself respectable'.
'Respectability' is, quite expressly, one of the major positives in her
vision of women's advance.[7]

At very much the same date Wollstonecraft's friend Mary Hays
was arguing for female rights with a similar emphasis on woman's
unilateral duty of sexual continence. She was a Unitarian, and one
line of descent of the feminist anti-sensual tradition is via W. J. Fox's
liberal Unitarian wing, represented by such writers as Mary Leman
Grimstone and Harriet Taylor. The latter's views I have already
cited. Grimstone was found congenial by the Owenite press, and
Owenite women not only tended to dissent from the libertarian programmes with which socialism toyed briefly in the 1830s, but in some
cases made explicit connections between the cause of women and a
moderate, orderly sexuality. The great Owenite lecturer Emma
Martin was particularly notable for her 'inverted puritanism' (as a
modern historian has termed it). Frances Morrison seems to have

[7] Wollstonecraft 1975: 87, 111, 114, 167.

been the solitary instance of an Owenite feminist who publicly embraced the prospect of a liberalized sexual code.[8]

Even Anna Wheeler, who has acquired a reputation for libertarian views, emerges in the Owenite press as entirely the heir of Mary Wollstonecraft in her belief that women's sexuality has been a 'fearful thing' which 'has fixed and perpetuated the degradation of her sex'. It is interesting that the main evidence for a more pro-sensual attitude in Wheeler comes from William Thompson's *Appeal of One Half of the Human Race* of 1825. Although Thompson credits Wheeler with a profound influence on this book, it may be that its protests against the barriers to women's 'gratification of her desires' are a case of the male-minded liberationism attacked by late-Victorian feminists and more recently by militants such as Sheila Jeffreys (in fact it is apparent from the text of the *Appeal* that Thompson is trying awkwardly to harmonize this aspect of his argument with a Wollstonecraftian belief that women's sexuality has been 'preposterously over-excited').[9]

It is important to stress how much more than a matter of strategy was the anti-sensual tone of nineteenth-century feminism: how much it reflected the actual *purpose* of women's emancipation as activists conceived this. An interesting case in point is the agitation for the admission of women to the medical profession. Ann Dally has recently uttered a sensible warning against generalizing about the motives of the women seeking to qualify as doctors from about the middle of Victoria's reign:

Some seriously sought control over their own lives and wanted satisfying outlets for their energy. Others had experience of sickness which influenced their choice . . . Many confessed to being simply ambitious . . . Some wanted to convince men they could do as well. Some sought social status . . . Some . . . sought to escape from romantic yearning . . . Some felt divinely inspired or were strongly moralistic and many shared the idea that women were morally superior to men. Others were fascinated by science.

But Dally, and other students of this subject, find that in practice the great bulk of the work performed by the new women doctors was gynaecological and obstetric. This tendency was doubtless driven largely by patient demand, and the existence of such a demand was the main overt argument of the campaign for women doctors, whatever the motives of individuals. The great ground on which women should

---

[8] Hays 1793: 84, 93; Robson 1987; *New Moral World*, 1 (1835), 132–5; B. Taylor 1983: 148; see p. 160 above.

[9] Pankhurst 1954; *Crisis*, 2 (1833), 279–81; W. Thompson 1825: pp. viii–xiii, 60–4, 77–8.

be admitted to the profession was the supposedly large number of female patients dismayed and deterred (rightly, it was said by the campaigners) in seeking medical help for distinctively female conditions by the fact that they must consult male practitioners. According to a recent historian of this agitation, it predated the campaigns for the admission of women to the civil service and the law, for example, because it was not about equality at all: 'the issue of women's entry to the medical profession was not seen by supporters as one of equal rights to employment . . . The crucial issue for them was the relieving of the suffering of women who were forced to consult male doctors'.

This emphasis is reflected in the story of attempts to provide medical qualifications for women (given that the ordinary teaching establishments quickly closed their gates against them). The first institution for this purpose in England was the Ladies' Medical College, founded in 1862. Despite its name it started as, and remained, an instrument for training middle-class women to be midwives: it may never have really had any higher ambitions. The cause of female midwifery was always dear to feminism: the leading example of a feminist promoting it both practically and theoretically is probably Emma Martin. The result was that feminism became an ally of the venerable opposition to male-midwives, which right through the eighteenth century had alleged the sexually corrupting effects of this kind of medical attendance for doctor and patient alike. This was medico-sexual moralism of a reductive and reactionary sort, in a quite different vein from feminist anti-sensualism, and the element of common ground between them is significant.

The next venture in women's medical education was the London School of Medicine for Women, set up in 1874. Its aim was consciously much more complete than that of the Ladies' Medical College: it sought to create women doctors qualified just as broadly and to the same standard as their male counterparts. But the school's woman co-founder, Sophia Jex-Blake—herself a famous pioneer medical entrant at Edinburgh—made it clear that the sole rationale for qualified female practitioners was that women could be treated by their own sex: 'women are the fit physicians for women, and men for men'. The school also insisted that certain of its students severed their connections with the newly founded, or rather refounded, Malthusian League in 1877.[10]

[10] Dally 1991: 200–4; Burstyn 1980: 85; Blake 1990: pp. xii, 79–83, 167–71; Donnison 1977: 73–7; Benn 1992: 116–31, 138–4; B. Taylor 1983: 155; Jex-Blake 1869.

For feminists, a future in which women had equal rights would
also be a less sensual world. An important part of their logic here—
which I do not wish to neglect—was the thought that women would
have many recognized roles in addition to those of fiancée-to-be,
fiancée, mistress, prostitute, wife, and mother. But the decay of the
sensual also featured in the feminist vision of an improved world simply in the way it featured in all progressive thought of the period: as
the corollary of a privileging of the intellect, and thus the non-
sensual, as the instigator and guide of social and political advance.
Mary Wollstonecraft's *Vindication* breathes the spirit of this
Enlightenment-derived reverence for the rational and mental aspect
of our human capacities on almost every page.

Even the looked-for enlargement of women's roles is foreseen in a
way that is characteristic of the culture: as involving an actual
modification of sexual function in response to new environmental
conditions. Mary Wollstonecraft believed that the current system
constrained women by 'cramping their understandings' (which we
would probably all accept as plausible), but also by 'sharpening their
senses'. Women whose only available roles were dependent on male
sexuality—who were 'sex objects', in our idiom—experienced a real
alteration in their own sexuality: 'they become the prey of their
senses, delicately termed sensibility, and are blown about by every
momentary gust of feeling'. In Wollstonecraft's account libido can be
positively altered in both directions—enhanced or reduced—
because its maximum level is not its natural level. Despite her overt
antagonism to Rousseau in the *Vindication* she agrees with him that
sexual drives are partly an artefact: 'civilised women are . . . so weak-
ened by false refinement, that, respecting morals, their condition is
much below what it would be were they left in a state nearer to
nature'.[11]

Such beliefs settled into a directly physical idiom with
Wollstonecraft's heirs. The well-exercised girl's body, taught
Elizabeth Blackwell, reaches puberty late, and a healthy woman's
sexual response in intercourse only becomes strong over time; sex is
actually 'even more a mental passion than a physical expression'.
Late-nineteenth-century feminists seem to have had less to say on
the artefactual character of women's sexual drives and responses, but
they had striking views on related functions, including one which we

---

[11] Wollstonecraft 1975: 104, 152.

nowadays would regard as intractably physical, and quite beyond the reach of cultural control. They argued for example that a woman's so-called 'maternal instincts' were an artefact (and one which obviously suited male purposes). Most notably, Elizabeth Wolstonecraft Elmy advanced the theory that menstruation was the product of inordinate male sexual demands, and would tend to disappear in a more conti-nent society.[12]

The modern reader is entitled to regard Elmy's menstrual doctrine as repressive and hare-brained—but only by his or her own lights. In the nineteenth-century context it is not especially eccentric, even though it is not reactionary in tendency either. Elmy believed in it because she was a feminist, not despite her feminism. And every-where we turn the progressive thought of the period is coloured by the wish, and confidence, that sex will become less and less impor-tant as human society advances. Elmy's theory is no odder than many moments in an odd, and misunderstood, story.

When I started to write this book five years ago feminist thought offered the best means of making the logic of Victorian anti-sensual-ism vivid, because still topical, to the modern reader. Since then sex between adults and children has become a notable subject of public concern—in Britain at least—and it is a phenomenon even better adapted to remind the reader that the sexual-moral spirit of the last century was not as alien as we may suppose. In so far as there is a very widespread and genuine repugnance for paedophilia, and agree-ment that it should be punished by the law, we are, I believe, in a very honorable sense 'Victorians' still.

This claim may seem strange and unwelcome. Many people per-ceive their disapproval of paedophilia as adequately grounded in util-itarian considerations, that is, in the suffering and psychological damage endured by children in sexual contacts with adults. The fact of such suffering and damage is often asserted in an a priori spirit, and evidently in ignorance of the large scientific literature investigat-ing the subject. This literature, while confirming that sexual abuse by adults can be terribly destructive, also shows that in some cases the suffering and damage caused is less clear-cut. Many opponents of paedophilia are, of course, aware that the issue is a complex one, which narrowly utilitarian arguments will not cope with, and they

---

[12] Blackwell 1859: 19–20, 1884: 15, 45–6; S. K. Kent 1987: 96–9, 110–11; Shanley 1989: 184–7; Jeffreys 1985: ch. 2.

rightly appeal to an additional consideration: that such acts are not adequately consented to by the children involved.

Do these arguments about suffering, damage, and lack of consent secure the degree of condemnation of paedophilia which those who trust in them would like to see? I doubt if they do. To start with, these arguments appear to be vulnerable in the borderline cases: where the children concerned are relatively mature or even pubescent. Rather subtle notions of suffering, damage, and non-consent may have to be invoked in such cases, and therefore more disputable ones. Our condemnation of paedophilia in these instances should be correspondingly prepared to admit exceptions. Secondly, suffering, damage, and non-consent in the victims of paedophilia cannot be absolute arguments against paedophiliac practices. They are effective arguments only if there are no considerations which outweigh them. Sexual pleasure can be claimed to be a good, and those who rely on these arguments open the door alarmingly to a defence of some paedophile acts on the grounds that the sexual pleasure of the adult outweighs the negative experiences of the child, or outweighs the child's lack of consent.

In suggesting that the utilitarian and consensual arguments are not strong enough I do not wish to be misunderstood. I regret the dependence which is put on them precisely because I believe paedophilia to be an evil. Those who agree with me are not giving their case good support, and probably not actually being faithful to their own moral intuitions, when they appeal only to arguments about outcomes and consent. If the condemnation of paedophilia is to be secure other considerations need to be invoked, or rather, I would suggest, brought into the light from the dark corners of our thinking to which they have been banished for their unfashionability.

One of these submerged ideas is that sexual pleasure and the capacity for sexual pleasure may not be a good, and in some contexts must be regarded as having no claims on us whatsoever. This idea, once accepted, would tend to justify the so-called chemical castration of paedophile adults, with or without their consent. I believe such a measure would be correct, drastic though it is, and not an illegitimate interference with an individual's sexuality. The present unsatisfactory standing of chemical castration, legally, is in my view a token of our embarrassment about granting that sexual pleasure may not be a good in a culture which likes to extol the supreme importance of the fulfilment of our sexual needs. It is this embarrassment, in fact, which drives people to an ultimately damaging dependence on arguments about outcomes and consent against paedophilia.

Chemical castration also bears closely on my main point here, which is that our repugnance for paedophilia is, in a good sense, 'Victorian'. The possibility that human libido could be reduced (by cultural rather than chemical mechanisms) was an empirical claim which was invoked quite readily in support of anti-sensual attitudes in the nineteenth century. Sensing a core of surviving anti-sensualism in our attitudes assists the act of historical imagination whereby we can sympathize with, for example, the Victorian resistance to libertarian sexual programmes. It may also assist us in holding positions on current sexual morality which we find desirable. Those who wish paedophilia to continue to be a crime cannot simply trust that this one area of sexual activity, almost uniquely, will be exempt from the liberalizing tendencies of our modern culture. It has not escaped the proponents of paedophilia that the arguments about outcomes and consent which are deployed against them are not strong, and that in a culture in which notions of sexual excess (and therefore also of restraint) have generally collapsed, there is a lack of other arguments which will be found respectable. Many paedophiliacs probably do genuinely regard our current restrictions on paedophiliac sex as a kind of fake, rather in the way individuals in the silent sub-culture of Victorian sexual hedonism regarded their era's predominant code of progressive anti-sensualism. There were signs of such an attitude among those who fomented hostility towards the two Cleveland paediatricians, Marietta Higgs and Geoffrey Wyatt, in 1987. Unless we take a leaf from the nineteenth-century anti-sensualist book, in other words, it is quite likely that in Britain the legal restraint on sex between children and adults will not survive.

Bibliography

ABRAMS, PHILIP, *The Origins of British Sociology* (Chicago, 1968).

ACTON, WILLIAM, *The Functions and Disorders of the Reproductive Organs* (1st edn. London, 1857a; 3rd edn. London, 1862; 4th edn. London, 1865; 6th edn. London, 1875).

—— *Prostitution* (1st edn. London, 1857b; 2nd edn. London, 1870).

ADAMS, WILLIAM BRIDGES, ('Junius Redivivus'), 'On the Condition of Women in England', *Monthly Repository*, 7 (1833), 217–31.

ADAMS, W. E., *Memoirs of a Social Atom* (London, 1903).

'AGATE, PETER' (? = E. Hoskyns), *Sexual Economy as Taught by Charles Bradlaugh* (London, 1885).

AINSLIE, ROBERT, *An Examination of Socialism* (London, 1840).

AINSWORTH, A. J., 'Religion in the Working Class Community and the Evolution of Socialism in Late Nineteenth Century Lancashire', *Histoire Sociale—Social History*, 10 (1977), 354–80.

ALLCHIN, A. M., *The Silent Rebellion* (London, 1958).

ALTHOLZ, JOSEPH L., 'Alexander Haldane, the *Record*, and Religious Journalism', *Victorian Periodicals Newsletter/Review*, 20 (1987), 23–31.

'Amicus Patriae', *Political and Moral Reflections during Twelve Rambles in London* (London, 1810).

ANDERSON, NANCY FIX, *Woman against Women in Victorian England* (Bloomington, Ind., 1987).

—— 'Eliza Lynn Linton, Dickens, and the Woman Question', *Victorian Periodicals Newsletter/Review*, 22 (1989), 134–41.

Anonyma (London, 1864).

ANSON, PETER F., *The Call of the Cloister* (Rev. edn. London, 1964).

ARMSTRONG, CHESTER, *Pilgrimage from Nenthead* (London, 1938).

ARMSTRONG, JOHN, 'Female Penitentiaries', *Quarterly Review*, 83 (1848), 359–76.

—— *Essays on Church Penitentiaries* (London, 1858).

ARMYTAGE, W. H. G., *Heavens Below* (London, 1961).

ARNDT, KARL J. R., *George Rapp's Harmony Society 1785–1847* (Philadelphia, 1965).

ASHTON, T. S., *Economic and Social Investigations in Manchester 1833–1933* (Manchester, 1934).

ASPINWALL, BERNARD, 'Social Catholicism and Health: Dr and Mrs Thomas Low Nichols in Britain', in W. J. Shiels, ed., *The Church and Healing* (Studies in Church History, 19; 1982), 249–70.

BADCOCK, JOHN, ('John Bee'), *Living Picture of London, for 1828* (London, 1828).

BAIN, ALEXANDER, *John Stuart Mill* (London, 1882).

BALLEINE, G. R., *Past Finding Out* (London, 1956).

BANKS, J.A. and OLIVE, 'The Bradlaugh–Besant Trial and the English Newspapers', *Population Studies*, 8 (1954), 22–34.

—— and —— *Feminism and Family Planning in Victorian England* (London, 1964).

BANKS, OLIVE, *Faces of Feminism* (Oxford, 1981).

BARKER, JOSEPH, *The Gospel Triumphant* (Newcastle, 1839).

—— *The Abominations of Socialism Exposed* (Newcastle, 1840*a*).

—— *The Overthrow of Infidel Socialism* (London, 1840*b*).

—— *The History and Confessions of a Man as Put forth by Himself* (Wortley, 1846).

—— *Confessions of Joseph Barker, a Convert from Christianity* (London, 1858).

—— *Love and Marriage* (London, 1863).

—— *Teachings of Experience* (London, 1869).

BAX, ERNEST BELFORT, *Reminiscences and Reflexions of a Mid and Late Victorian* (London, 1918).

BAYLEY, J., *Mr Roebuck's Want of Principle in his Animadversions Exposed and Condemned* (Accrington, 1838*a*).

—— *Remarks on Mr Roebuck's Admissions of Error* (Manchester, 1838*b*).

BEALE, LIONEL S., *Our Morality and the Moral Question* (London, 1887).

BEARD, JOHN RELLY, *The Religion of Jesus Christ Defended from the Assaults of Owenism* (London, 1839).

BEBBINGTON, D. W., *Evangelicalism in Modern Britain* (London, 1989).

BEDDOES, THOMAS, *Hygeia or Essays Moral and Medical* (Bristol, 1802).

BELL, R., *Religion and Morality Vindicated* (London, 1813).

BENN, J. MIRIAM, *The Predicaments of Love* (London, 1992).

BENTHAM, JEREMY, *The Works of Jeremy Bentham*, vol. i, ed. John Bowring (London, 1843).

BESANT, ANNIE, *The Law of Population* (London, 1879).

—— *Henry Varley Exposed* (London, 1881).

—— *Marriage* (2nd edn. London, 1882).

—— *Autobiographical Sketches* (London, 1885).

—— *An Autobiography* (London, 1893).

BEST, G. F. A., 'Popular Protestantism in Victorian Britain', in Robert Robson, ed., *Ideas and Institutions of Victorian Britain* (London, 1967), 115–42.

BESTOR, ARTHUR EUGENE, *Backwoods Utopias* (Philadelphia, 1950).

BEVAN, WILLIAM, *Prostitution in the Borough of Liverpool* (Liverpool, 1843).

BEVINGTON, MERLE MOWBRAY, *The Saturday Review 1855–1868* (New York, 1941).

BICKERSTETH, E. (ed.), *Prize Essays on the Temporal Advantages of the Sabbath to the Labouring Classes* (London, 1849).

BIRCH, EDWARD, *Remarks on Socialism* (London, 1839).

BLACKBURN, Revd J., 'The Divine Institution and Sacred Obligation of Marriage', *Christian Instruction Society: A Course of Thirteen Lectures to Socialists* (London, 1840), 95–100.

BLACKMORE, JOHN, *Midnight Cruizes* (London, 1854).

—— *The London by Moonlight Mission* (London, 1860).

—— *Cruisings* (London, 1889).

BLACKWELL, ELIZABETH, *The Laws of Life* (London, 1859).

—— *The Human Element in Sex* (London, 1884).

—— *The Influence of Women in the Profession of Medicine* (London, 1889).

BLAKE, CATRIONA, *The Charge of the Parasols* (London, 1990).

BLANCO WHITE, JOSEPH, *Practical and Internal Evidence against Catholicism* (London, 1825).

BLANCHARD, E. L., *The Life and Reminiscences of E. L. Blanchard*, ed. Clement Scott and Cecil Howard (London, 1891).

BLAND, LUCY, 'The Married Woman, the "New Woman" and the Feminist: Sexual Politics of the 1890s', in Jane Rendall, ed., *Equal or Different* (Oxford, 1987), 141–64.

BLATCHFORD, ROBERT, *My Eighty Years* (London, 1931).

BLUNT, JOHN HENRY, *Dictionary of Sects* (London, 1874).

BOASE, GEORGE C., 'The Lord Chief Baron Nicholson', *Notes and Queries*, 8th ser. 3 (1893), 3–5.

BONER, HAROLD A., *Hungry Generations* (New York, 1955).

BONHAM, VALERIE, *A Joyous Service* (Windsor, 1989).

BONNER, HYPATIA BRADLAUGH, *Charles Bradlaugh* (London, 1895).

BOURNE, STEPHEN, 'On the Growth of Population with Relation to the Means of Sustenance', *Report of the Forty-Seventh Meeting of the British Association for the Advancement of Science* (1878), 165–73.

BOWES, JOHN, *The Social Beasts* (London, 1840).

—— and JONES, LLOYD, *Socialism. Report of a Public Discussion between John Bowes, Minister of the Gospel, Liverpool, and Lloyd Jones, Social Missionary, Glasgow* (London, 1840a).

—— and —— *Report of the Discussion on Marriage, as Advocated by Robert Owen, between L. Jones and J. Bowes* (Liverpool, 1840b).

BRADLAUGH, CHARLES, *Jesus, Shelley and Malthus* (London, 1861).

BRADLEY, IAN, *The Call to Seriousness* (London, 1976).

BRAY, JOHN FRANCIS, *Labour's Wrongs and Labour's Remedy* (Leeds, 1839).

BRIERLEY, BEN, *Home Memories* (Manchester, 1887).

BRIGGS, ASA, *Victorian People* (Harmondsworth, 1965).

BRINDLEY, JOHN, *The Immoralities of Socialism* (Birmingham, 1840).

—— and HOLLICK, F. *Public Discussion . . . between Mr Brindley and Mr Hollick* (Leicester, 1840).

—— and WOODMAN, WOODVILLE, *The Teachings of Swedenborg on Marriage . . . a Debate between Dr Brindley and the Rev. Woodville Woodman* (London, 1862).

BRISTOW, EDWARD J., *Vice and Vigilance* (Dublin, 1977).

BROWN, CALLUM G., 'Did Urbanisation Secularize Britain?', *Urban History Yearbook* (1988), 1–14.

BROWN, FORD K., *Fathers of the Victorians* (Cambridge, 1961).

BRUCE, W., *Marriage* (London, 1850).

BUCHANAN, ROBERT, *Socialism Vindicated* (Manchester, 1840).

—— *An Exposure of the Falsehoods, Calumnies, and Misrepresentations of a Pamphlet Entitled 'The Abominations of Socialism Exposed'* (Manchester, 1841).

BUDD, SUSAN, *Varieties of Unbelief* (London, 1977).

BURN, JAMES DAWSON, *The Autobiography of a Beggar Boy* (London, 1855).

BURSTYN, JOAN N., *Victorian Education and the Ideal of Womanhood* (London, 1980).

BUTLER, JOSEPHINE, 'Lovers of the Lost', *Contemporary Review*, 13 (1870), 16–40.

—— *An Autobiographical Memoir* (Bristol, 1928).

CAMPBELL, ALEXANDER, and BANNISTER, J. T., *Socialism, Public Discussion between Mr Alexander Campbell and Rev. J. T. Bannister* (Coventry, 1839).

—— and ROBERTS, W. P., *Authentic Report of the Discussion at the Guildhall, Bath . . . between Mr Alexander Campbell, Social Missionary; and Mr W. P. Roberts* (Bath, 1838).

CARLILE, RICHARD, *Every Woman's Book* (4th edn. London, 1826).

CARLYLE, THOMAS, *Critical and Miscellaneous Essays* (London, 1899).

CARTER, THOMAS, *Memoirs of a Working Man* (London, 1845).

CARTER, THELLUSSON, THOMAS, *Is it Well to Institute Sisterhoods in the Church of England for the Care of Female penitents?* (London, 1853).

—— *The First Ten Years of Clewer* (London, 1861).

CHAMBERS, ROBERT, *Vestiges of Creation* (London, 1844).

CHANCELLOR, EDWIN BERESFORD, *Annals of the Strand* (London, 1912).

CHANDRASEKHAR, S., 'A Dirty, Filthy Book' (Berkeley, Calif., 1981).

CHANNING, WILLIAM, *Correspondence of William Ellery Channing DD and Lucy Aikin*, ed. Anna le Breton (London, 1874).

CHARLES, A. O., *The Female Mission to the Fallen* (London, 1860).

CHITTY, SUSAN, *The Beast and the Monk* (London, 1974).

CHRISTIE, O. F., *The Transition from the Aristocracy* (London, 1927).

CHURCH, JOHN, *The Believer's Confidence in a Faithful God* (London, 1817).

—— *The Voice of Faith in the Valley of Achor* (London, 1818).

—— *The Gracious Designs of God Accomplished* (London, 1819).

CHURCHILL, THOMAS F., *The Substance of a Sermon . . . Occasioned by the Sudden Demise of Mary Minting* (London, 1818).

CLARK, DAVID LEE (ed.), *Shelley's Prose* (London, 1988).

CLARKE, M. L., *George Grote* (London, 1962).

'A Clerical Gentleman', *A Concise and Convincing Argument against Socialism* (London, 1840).

CLODD, EDWARD, *Memories* (London, 1916).

CLOWES, JOHN, *The Golden Wedding Ring* (London, 1888).

COLERIDGE, SAMUEL TAYLOR, *Biographia Literaria*, ed. James Engell and Walter Jackson Bate (Princeton, NJ, 1983).

COLLINS, PHILIP, *Dickens and Crime* (London, 1964).

COMINOS, PETER T., 'Late-Victorian Sexual Respectability and the Social System', *International Review of Social History*, 8 (1963), 18–48, 216–50.

CONYBEARE, W. J., 'Church Parties', *Edinburgh Review*, 98 (1853), 273–342.

CONWAY, MONCURE D., *Liberty and Morality* (London, 1878).

COOKE TAYLOR, W., 'The Employment of Married Women in Manufacture', *Transactions of the National Association for the Promotion of Social Science* (1874), 605–13.

COOPER, THOMAS, *The Life of Thomas Cooper* (Leicester, 1971).

COTT, NANCY F., 'Passionlessness: An Interpretation of Victorian Sexual Ideology, 1790–1850', *Signs*, 4 (1978), 219–36.

COUPLAND, W. C., *The Principle of Individual Liberty, How Far Applicable to the Relations of the Sexes* (London, 1878).

CROMPTON, LOUIS, *Byron and Greek Love* (London, 1985).

CURRIE, ROBERT, GILBERT, ALAN, and HORSLEY, LEE, *Churches and Churchgoers* (Oxford, 1977).

DALLY, ANN, *Women under the Knife* (London, 1991).

DALTON, T., and JONES, LLOYD, *Socialism Examined, Report of a Public Discussion . . . between Rev. T. Dalton . . . and Mr Lloyd Jones* (Manchester, 1838).

D'ARCY, F., 'The Malthusian League and the Resistance to Birth Control Propaganda in late Victorian England', *PS* 31 (1977), 429–48.

DAVIES, C. MAURICE, *Unorthodox London* (London, 1873).

—— *Heterodox London* (2nd edn. London, 1874).

—— *Orthodox London 2nd Series* (London, 1875).

DECK, J. G., *A Word of Warning* (London, 1845).

DICKENS, CHARLES, 'Home for Homeless Women', *Household Words*, 7 (1853), 161–75.

—— *The Letters of Charles Dickens*, vol. iv, ed. Kathleen Tillotson (Oxford, 1977).

—— *The Letters of Charles Dickens*, vol. v, ed. Graham Storey and K. J. Fielding (Oxford, 1981).

—— *The Letters of Charles Dickens*, vol. vi, ed. Graham Storey, Kathleen Tillotson, and Nina Burgis (Oxford, 1988).

DIXON, W. HEPWORTH, *Spiritual Wives* (London, 1868).

DOLBY, THOMAS, *Memoirs* (London, 1827).

DOLLIMORE, ARNOLD, *The Life of Edward Irving* (Edinburgh, 1983).

DONNISON, JEAN, *Midwives and Medical Men* (London, 1977).

DORÉ, GUSTAVE, and JERROLD, WILLIAM BLANCHARD, *London: A Pilgrimage* (London, 1872).

DRYSDALE, CHARLES, 'Memoir of the Author', in George Drysdale, *The Elements of Social Science* (35th edn. London, 1905), pp. i–vi.

DRYSDALE, GEORGE, *The Elements of Social Science* (5th edn. London, 1868).

DUBOIS, ELLEN CAROL, and GORDON, LINDA, 'Seeking Ecstasy on the Battlefield: Danger and Pleasure in Nineteenth-Century Feminist Thought', in Carole S. Vance, ed., *Pleasure and Danger* (Boston, Mass., 1984, 31–49.

DYMOND, JONATHAN, *Essays on the Principles of Morality* (London, 1829).

ELLEGARD, ALVAR, 'The Readership of the Periodical Press in Mid-Victorian Britain II', *Victorian Periodicals Newsletter/Review*, 13 (Sept. 1971), 3–22.

ENAULT, LOUIS, *Londres* (Paris, 1876).

ENGLANDER, DAVID, 'The Word and the World: Evangelicalism in the Victorian City', in Gerald Parsons, ed., *Religion in Victorian Britain* (Manchester, 1988), ii. 14–38.

EVANS, ERIC J., *The Forging of the Modern State* (London, 1983).

EVANS, JOHN, *A Letter Addressed to Robert Hawker* (London, 1809).

'Examiner of the New Church Doctrines', *Refutation of Mr J. H. Roebuck's Pamphlet* (London, 1839).

EYRE, C. S., *A Few Words on Socialism* (Coventry, 1840).

Fast Life (London, 1859).

FAUCHER, LEON, *Manchester in 1844* (London, 1844).

FAWCETT, HENRY, *Pauperism: Its Causes and Remedies* (London, 1871).

FELLOWES, ROBERT, *Religion without Cant* (London, 1801).

Female Mission to the Fallen, *Reports* (1866–8).

FIELD, JAMES ALFRED, *Essays on Population* (Chicago, 1931).

FIELDEN, THOMAS, *An Exposition of the Fallacies and Absurdities of that Deluded Church Generally Known as Christian Israelites* (Rawtenstall, 1856).

FINNEGAN, FRANCES, *Poverty and Prostitution* (Cambridge, 1979).

FISKE, STEPHEN, *English Photographs by an American* (London, 1869).

FOLBRE, NANCY, '"The Improper Arts": Sex in Classical Political Economy', *Population and Development Review*, 18 (1992), 105–21.

'Fontana and Prati', *St Simonism in London* (London, 1833).

FONTANE, THEODOR, *Journeys to England in Victoria's Early Days* (London, 1939).

FOOTE, G. W., *The Hall of Science Libel Case* (London, 1895).

FOURIER, CHARLES, *The Passions of the Human Soul*, trans. John Reynell Morell (London, 1851).

FOX, W., *The Secret of the Lord in Judgement* (Reading, 1886).

FOX, WILLIAM JOHNSON, 'The Dissenting Marriage Question', *Monthly Repository*, 7 (1833a), 136–42.

—— 'A Victim', *Monthly Repository*, 7 (1833b), 164–77.

—— 'Campbell's Life of Mrs Siddons', *Monthly Repository*, 8 (1834), 533–50.

FROST, THOMAS, *Forty Years' Recollections* (London, 1880).

FRYER, PETER, *The Birth Controllers* (London, 1965).

GAMOND, Madame GATTI DE, *The Phalanstery* (London, 1841).

GARNETT, RICHARD (with Edward Garnett), *The Life of W. J. Fox* (London, 1910).

GAY, PETER, *The Bourgeois Experience*, i. *Education of the Senses* (Oxford, 1984).

GILBERT, ALAN D., *Religion and Society in Industrial England* (Harlow, 1976).

GILES, JOHN EUSTACE, *Socialism as a Religious Theory, Irrational and Absurd* (London, 1839a).

—— *Socialism, in its Moral Tendencies, Compared with Christianity* (London, 1839b).

GILLESPIE, WILLIAM, *Atheism or Theism?* (London, 1869–72).

GISBORNE, THOMAS, *An Enquiry into the Duties of the Female Sex* (London, 1797).

GLASS, DAVID V., *Population Policies and Movements in Europe* (London, 1940).

GOODELL, WILLIAM, *Lessons in Gynecology* (Philadelphia, 1879).

GOODMAN, MARGARET, *Experiences of an English Sister of Mercy* (London, 1862).

GOULD, F. J., *The Life-Story of a Humanist* (London, 1923).

GOYDER, D. G., *The Principles and Tendency of Swedenborgianism Considered and Defended* (Glasgow, 1839).

GOYDER, D. G., *The Autobiography of a Phrenologist* (London, 1857).

GRANT, JAMES, *Portraits of Public Characters* (London, 1841).

GREAVES, JAMES PIERREPONT, *Letters and Extracts from the Ms Writings of James Pierrepont Greaves*, ed. Alexander Campbell (Ham Common, 1843).

GREENWOOD, JAMES, *The Wilds of London* (London, 1874).

GREG, WILLIAM RATHBONE, 'Prostitution', *Westminster Review* 53 (1850), 448–506.

—— *The Creed of Christendom* (London, 1851).

GROTE, HARRIET, *Collected Papers* (London, 1862).

Guardian Society, *Reports* (London, 1816–45).

GUTTERIDGE, JOSEPH, *Master and Artisan in Victorian England*, ed. Valerie E. Chancellor (London, 1969).

HAIG, ALAN, *The Victorian Clergy* (Beckenham, 1984).

HALE, WILLIAM, *An Address to the Public upon the Dangerous Tendency of the London Female Penitentiary* (London, 1809).

—— *Considerations of the Causes and the Prevalence of Female Prostitution* (London, 1812).

HALEVY, ELIE, *A History of the English People in 1815* (London, 1924).

HALL, CATHERINE, 'The Early Formation of Victorian Domestic Ideology', in Sandra Burman, ed., *Fit Work for Women* (London, 1979), 15–32.

HALL, THEODORE, *The Gordian Knot Untied* (Manchester, 1841).

HAMMERTON, A. JAMES, *Cruelty and Companionship* (London, 1992).

HANCOCK, EDWARD, *A Scourge for Deists and Atheists* (London, 1836).

—— *Robert Owen's Community System* London, 1837).

HANGER, GEORGE, *The Life, Adventures, and Opinions of Col. George Hanger* (London, 1801).

HARDY, DENNIS, *Alternative Communities in Nineteenth Century England* (London, 1979).

HARDY, SHEILA, *The Diary of a Suffolk Farmer's Wife, 1854–1869* (Basingstoke, 1992).

HARDY, THOMAS, *Memoir of Thomas Hardy* (London, 1832).

HARRISON, AUSTIN, *Frederic Harrison: Thoughts and Memories* (London, 1926).

HARRISON, BRIAN, *Drink and the Victorians* (London, 1971).

—— 'Pubs', in H. J. Dyos and Michael Wolff, eds., *The Victorian City* (London, 1973), i. 161–90.

—— *Peaceable Kingdom* (Oxford, 1982).

—— and HOLLIS, PATRICIA, 'Chartism, Liberalism and the Life of Robert Lowery', *Past and Present*, 82 (1967), 503–35.

—— and —— *Robert Lowery Radical and Chartist* (London, 1979).

HARRISON, J. F. C., *Robert Owen and the Owenites* (London, 1969).

—— *The Second Coming* (London, 1979).

HARTE, RICHARD, *On the Laws and Customs Relating to Marriage* (London, 1870).

HAUGHTON, G. D., *On Sex in the World to Come* (London, 1841).

HAYS, MARY, *Letters and Essays, Moral, and Miscellaneous* (London, 1793).

HEASMAN, KATHLEEN, *Evangelicals in Action* (London, 1962).

HEATH, WILLIAM, *Paul Pry's Ramble through the 'New Moral World'* (Doncaster, 1838).

HEENEY, BRIAN, *A Different Kind of Gentleman* (Hamden, Conn., 1976).

HELSINGER, ELIZABETH K., SHEETS, ROBIN LAUTERBACH, and VEEDER, WILLIAM, *The Woman Question* (Chicago, 1983).

HIBBERT, H. G., *Fifty Years of a Londoner's Life* (London, 1916).

HILL, MICHAEL, *The Religious Order* (London, 1973).

HIMES, NORMAN, 'The Place of John Stuart Mill and of Robert Owen in the History of English Neo-Malthusianism', *Quarterly Journal of Economics*, 42 (1928), 627–40.

—— 'John Stuart Mill's Attitude towards Neo-Malthusianism', *Economic History*, 1 (1929a), 457–84.

—— 'McCulloch's Relation to the Neo-Malthusian Propaganda of his Time', *Journal of Political Economy*, 34 (1929b), 73–86.

—— 'Robert Dale Owen, the Pioneer of American Neo-Malthusianism', *American Journal of Sociology*, 35 (1930a), 529–47.

—— *Place on Population* (London, 1930b).

—— *Medical History of Contraception* (London, 1936).

HINDMARSH, ROBERT, *A Vindication of the Character and Writings of the Hon. Emmanuel Swedenborg* (Manchester, 1822a).

—— *Letter to Dr Priestley* (2nd edn. London, 1822b).

Hints to the Public and the Legislature on the Prevalence of Vice (London, 1811).

HITHERSAY, RICHARD B., and ERNEST, GEORGE, *Sketch of the Life of Saladin* (London, 1887).

HOBSON, JOSHUA, *Socialism as it is!* (Leeds, 1838).

HOLCOMBE, WILLIAM H., *The Sexes Here and Hereafter* (London, 1905).

HOLINSWORTH, C. B. (ed.), *Zion's Works* (London, 1899–1901).

HOLLINGSHEAD, JOHN, *My Lifetime* (London, 1895).

HOLLIS, PATRICIA, *The Pauper Press* (London, 1970).

HOLYOAKE, AUSTIN, *Large or Small Families?* (London, 1870).

HOLYOAKE, GEORGE JACOB ('Quasimodo'), *Joseph Barker and his Expulsion from the Secular Body* (London, 1861).

—— *John Stuart Mill as some of the Working Classes Knew him* (London, 1873).

—— *The History of Cooperation in England* (London, 1875).

—— *The Warpath of Opinion: Strange Things Seen Thereon* (London, 1896).

HOOD, EDWIN PAXTON, *The Earnest Minister* (London, 1856).

HOPKINS, ELLICE, *Work in Brighton* (London, 1877).

—— *Notes on Penitentiary Work* (London, 1879).

HOPKINS, JAMES K., *A Woman to Deliver her People* (Austin, Tex., 1982).

HOPKINSON, JAMES, *Victorian Cabinet Maker*, ed. J. B. Goodman (London, 1968).

HORSTMAN, ALLEN, *Victorian Divorce* (London, 1985).

HOUSMAN, LAURENCE, *The Unexpected Years* (London, 1937).

HOWSE, ERNEST MARSHALL, *Saints in Politics* (London, 1953).

HUDSON, DEREK, *Munby Man of Two Worlds* (London, 1974).

HUDSON, T. H., *Christian Socialism, Explained and Enforced, and Compared with Infidel Fellowship* (London, 1839).

HUNT, LEIGH, *An Attempt to Show the Folly and Danger of Methodism* (London, 1809).

—— *The Religion of the Heart* (London, 1853).

HUNTER, H. J., 'Inquiry on the State of the Dwellings of Rural Labourers', PP 1865, xxvi. 126–302.

HUXLEY, ALDOUS, *The Olive Tree* (London, 1936).

HYDE, J., *Character its Elements and Development* (London, 1875).

HYLSON-SMITH, KENNETH, *Evangelicals in the Church of England 1734–1984* (Edinburgh, 1988).

INGLIS, K. S., *Churches and the Working Classes in Victorian England* (London, 1963).

An Interesting Account of an Inmate of the Liverpool Penitentiary (London, 1835).

JAEGER, MURIEL, *Before Victoria* (London, 1956).

JAMES, PATRICIA, *Population Malthus* (London, 1979).

JAY, ELIZABETH, *The Religion of the Heart* (Oxford, 1979).

JAY, HARRIET, *Robert Buchanan* (London, 1903).

JEFFREYS, SHEILA, *The Spinster and her Enemies* (London, 1985).

—— *Anticlimax* (London, 1990).

JENKINS, JOHN EDWARD, 'Two Solutions', *Frasers*, NS 3 (1871), 451–6.

JEX-BLAKE, SOPHIA, 'Medicine as a Profession for Women', in Josephine Butler, ed., *Woman's Work and Woman's Culture* (London, 1869), 78–120.

JOHNSON, WENDELL STACY, *Living in Sin: the Victorian Sexual Revolution* (Chicago, 1979).

JONES, LLOYD, *The Life, Times and Labours of Robert Owen* (London, 1895).

JONES, OWAIN W., 'Prince and the Lampeter Brethren', *Trivium*, 5 (1970), 10–20.

JUDGE, HENRY, *Our Fallen Sister* (London, 1874).

'Juvenis', *Cursory Remarks on a Recent Publication* (London, 1809).

KENT, JOHN, 'Feelings and Festivals: An Interpretation of some Working-class Religious Attitudes', in H. J. Dyos and Michael Wolff, eds., *The Victorian City* (London, 1973), ii. 855–71.

KENT, SUSAN KINGSLEY, *Sex and Suffrage in Britain, 1860–1914* (Princeton, NJ, 1987).

KERVIGAN, AURELE, *L'Angleterre telle qu'elle est* (Paris, 1860).

KIDD, WILLIAM, *London and All its Dangers* (London, 1837).

KIRK, NEVILLE, *The Growth of Working-Class Reformism in Mid-Victorian Britain* (London, 1985).

KITCHENER, HENRY THOMAS, *Letters on Marriage* (London, 1812).

KNOWLTON, CHARLES, *Fruits of Philosophy* (London, 1877).

KNOX, ROBERT, ('A Physician'), *The Greatest of our Social Evils: Prostitution* (London, 1857).

KNOX, RONALD A., *Enthusiasm* (Oxford, 1950).

KODITSCHEK, THEODORE, *Class Formation and Urban Industrial Society: Bradford 1750–1850* (Cambridge, 1990).

KRUPPA, PATRICIA S., '"More Sweet and Liquid than any Other": Victorian Images of Mary Magdalene', in R. W. Davis and R. J. Helmstader, eds., *Religion and Irreligion in Victorian Society* (London, 1992), 117–32.

LACKINGTON, JAMES, *Memoirs of the First Forty-Five Years of the Life of J.L.* (1st edn. London, 1791; 13th edn. London, 1810).

—— *The Confessions of James Lackington* (London, 1804).

LANGER, WILLIAM L., 'The Origins of the Birth Control Movement in England in the Nineteenth Century', *Journal of Interdisciplinary History*, 5 (1975), 669–86.

LANSBURY, GEORGE, *My Life* (London, 1928).

LAQUEUR, THOMAS W., 'Working-Class Demand and the Growth of English Elementary Education, 1750–1850', in Lawrence Stone, ed., *Schooling and Society* (Baltimore, 1976), 192–205.

LA TOUCHE, JOHN, *The Sins and Sorrows of London* (London, 1862).

LAWSON, JOSEPH, *Letters to the Young on Progress in Pudsey during the Last Sixty Years* (Stanningley, 1887).

LEA, JOHN WALTER, *Christian Marriage: Its Open and Secret Enemies* (London, 1881).

LEDBETTER, ROSANNA, *A History of the Malthusian League 1877–1928* (Columbus, Oh., 1976).

LEOPOLD, RICHARD WILLIAM, *Robert Dale Owen* (Cambridge, 1940).

LE PLAY, M. F., *Les Ouvriers Européens* (Paris, 1855).

'Le Plus Bas', *Our Plague Spot: In Connection with our Polity and Usages* (London, 1859).

L'ESPERANCE, JEAN LAWRENCE, 'Woman's Mission to Woman: Explorations in the Operation of the Double Standard in Nineteenth Century England', *Histoire Sociale—Social History*, 12 (1979), 316–38.

LEVENTHAL, F. M., *Respectable Radical* (London, 1971).

LEVINE, PHILIPPA, *Victorian Feminism 1850–1900* (London, 1987).

—— *Feminist Lives in Victorian England* (Oxford, 1990).

LEWIS, DONALD, *Lighten their Darkness* (New York, 1986).

LINDRIDGE, JAMES, *The Merry Wives of London* (London, 1850).

LINEHAM, PETER JAMES, *The English Swedenborgians 1770–1840*, Univ. of Sussex, Ph.D. thesis (1978).

LOGAN, WILLIAM, *An Exposure, from Personal Observation, of Female Prostitution in London, Leeds, and Rochdale, and Especially in the City of Glasgow* (2nd edn. Glasgow, 1843).

London by Night (London, ?1857).

London Female Mission, *Reports* (1838–9).

London Female Penitentiary, *Reports* (1808–42).

The London Guide (London, 1818).

London Society for the Protection of Young Females, *Fourth Report* (London, 1839).

LOVETT, WILLIAM, *Social and Political Morality* (London, 1853).

—— *The Life and Struggles of William Lovett* (London, 1876).

MCCABE, JOSEPH, *Life and Letters of George Jacob Holyoake* (London, 1908).

MCCALMAN, IAIN, *Radical Underworld* (Cambridge, 1988).

—— 'Popular Irreligion in Early Victorian England: Infidel Preachers and Radical Theatricality in 1830s London', in R. W. Davis and R. J. Helmstader, eds., *Religion and Irreligion in Victorian Society* (London, 1992), 51–67.

MCCARTHY, JUSTIN, 'Novels with a Purpose', *Westminster Review*, NS 26 (1864), 24–49.

MCCORMICK, DONALD, *Temple of Love* (London, 1962).

MACKAY, CHARLES R., *Life of Charles Bradlaugh M.P.* (London, 1888).

MCLAREN, ANGUS, *Birth Control in Nineteenth-century England* (London, 1978).

MCLEOD, ALEXANDER, *Christus Consolator* (London, 1870).

MCLEOD, HUGH, 'Class, Community and Region: The Religious Geography of Nineteenth-Century England', *Sociological Yearbook of Religion in England*, 6 (1973), 29–72.

—— *Class and Religion in the Late Victorian City* (London, 1974).

—— *Religion and the Working Class in Nineteenth-Century Britain* (London, 1984).

MADAN, MARTIN, *Thalyphthora* (2nd edn. London, 1781).

MALLOCK, W. H., 'Marriage and Free Thought', *Fortnightly Review*, 50 (1891), 259–78.

MALMGREEN, GAIL, *Silk Town: Industry and Culture in Macclesfield 1750–1835* (Hull, 1985).

—— *Religion in the Lives of English Women 1760–1930* (London, 1986).

MANDER, CHARLES, *The Reverend Prince and his Abode of Love* (East Ardsley, 1976).

MANNING, HENRY EDWARD, *Penitents and Saints* (London, 1844).

MARWICK, W. H., *The Life of Alexander Campbell* (Glasgow, 1964).

MASON, MICHAEL, *The Making of Victorian Sexuality* (Oxford, 1994).

MATHER, JOSEPH, *Socialism Exposed* (London, 1839).

MATTHEWS, ROBERT, 'Is Marriage Worth Perpetuating?', *Lectures against Socialism* (London, 1840), no. 9.

MATTHEWS, RONALD, *English Messiahs* (London, 1936).

MAYHEW, HENRY, *London Labour and the London Poor* (London, 1861–2).

MAYHEW, HORACE, 'The Husband of the Strong-Minded Woman', *British Journal*, 3 (1853), 290–3.

MENDUS, SUSAN, 'The Marriage of True Minds: The Ideal of Marriage in the Philosophy of John Stuart Mill', in Susan Mendus and Jane Rendall, eds., *Sexuality and Subordination* (London, 1989), 171–91.

Metropolitan Female Asylum, *Report of the Provisional Committee* (London, 1830).

'M.G.H.', *Poverty: Its Cause and Cure* (London, 1870).

MICKLEWRIGHT, F. H. AMPHLETT, 'The Rise and Decline of English Neo-Malthusianism', *Population Studies* 15 (1962), 32–51.

MILES, DUDLEY, *Francis Place* (Brighton, 1988).

MILL, JAMES, *The Article Colony* (London, 1820).

MILL, JOHN STUART, *The Early Draft of John Stuart Mill's Autobiography*, ed. Jack Stillinger (Urbana, Ill., 1961).

—— *The Earlier Letters of John Stuart Mill*, ed. Francis Mineka (Toronto, 1963).

—— *The Later Letters of John Stuart Mill 1849–1873*, ed. Francis Mineka and Dwight D. Linbley (Toronto, 1972).

—— *Essays on Equality, Law and Education*, ed. John M. Robson and Stefan Collini (Toronto, 1984).

—— *Autobiography* (London, 1989).

MINEKA, FRANCIS E., *The Dissidence of Dissent* (Chapel Hill, NC, 1944).

Missionary and Tract Society of the New Jerusalem Church, *Tracts on Various Subjects* (London, 1844).

MITCHELL, DAVID, *The Fighting Pankhursts* (London, 1967).

MITCHELL, GEOFFREY (ed.), *The Hard Way Up* (London, 1968).

MITCHELL, SALLY, 'Sentiment and Suffering: Women's Recreational Reading in the 1860s', *Victorian Studies*, 21 (1977), 29–45.

MONTGOMERY, JOHN, *Abodes of Love* (London, 1962).

MORE, HANNAH, *Works* (London, 1830).

—— *Works* (London, 1833–4).

MORLEY, JOHN, 'Marriage in the Nineteenth Century', *Dublin University Magazine*, 59 (1862), 542–53.

—— 'A Short Letter to Some Ladies', *Fortnightly Review*, NS 7 (1870), 372–6.

MORRIS, R. J., *Class, Sect and Party* (Manchester, 1990).

MORRISON, C., *An Essay on the Relations between Labour and Capital* (London, 1854).

MORRISON, FRANCES, *The Influence of the Present Marriage System upon the Character and Interests of Females Contrasted with that Proposed by Robert Owen Esq* (Manchester, 1840).

MORT, FRANK, 'Purity, Feminism, and the State: Sexuality and Moral Politics, 1880–1914', in Mary Langrand and Bill Schwarz, eds., *Crises in the British State 1880–1930* (London, 1985), 209–25.

Mr Prince and the Agapemone (Taunton, 1858).

MULOCK, DIANA, *A Woman's Thoughts about Women* (London, 1858).

MUNCY, RAYMOND LEE, *Sex and Marriage in Utopian Communities* (Bloomington, Ind., 1973).

My Secret Life (New York, 1966).

National Community Friendly Society, *Social Tracts* (London, 1838–40).

NEAD, LYNDA, *Myths of Sexuality* (Oxford, 1988).

NETHERCOT, A. H., *The First Five Lives of Annie Besant* (London, 1961).

NEWMAN, FRANCIS W., 'F. W. Newman on the Population Question', *Barker's Review*, 1 (1861), 3–6.

—— 'Marriage Laws', *Frasers*, 76 (1867), 167–89.

—— *The Cure of the Great Social Evil* (London, 1869).

—— 'Malthusianism, True and False', *Frasers*, NS 3 (1871), 584–98.

—— 'What is Prostitution?', *Miscellanies*, iii (London, 1889), 285–8.

—— and BLACKWELL, ELIZABETH, *The Corruption Called Neo-Malthusianism* (London, 1889).

NEWSOME, DAVID, *The Parting of Friends* (London, 1966).

NICHOLS, T. L., *Human Physiology the Basis of Sanitary and Social Science* (London, 1872).

—— *Esoteric Anthropology* (London, 1873).

—— *Marriage in all Ages and Nations* (London, 1886).

NICHOLSON, RENTON, *Rogue's Progress*, ed. John L. Bradley (London, 1965).

NOBLE, SAMUEL, *An Appeal in behalf of the Views of the Eternal World and State* (London, 1838).

O Bit ov o Chat obeeat teese Lecturs un Things (Haslingden, 1844).

ODHNER, CARL T., *Annals of the New Church* (Bryn Athyn, 1904).

O'DONNOGHUE, H. C., *Marriage; The Source, Stability and Perfection of Human Happiness and Duty* (2nd edn. London, 1836).

OKEY, THOMAS, *A Basketful of Memories* (London, 1930).

OWEN, ROBERT, *Lectures on an Entire New State of Society* (London, 1830).

—— *The Social Bible* (with *Social Hymns*) (Manchester, 1835).

—— *Six Lectures Delivered in Manchester* (Manchester, 1837).

—— *The Marriage System of the New Moral World* (Leeds, 1839).

OWEN, ROBERT DALE, *Situations* (London, 1840).

—— *Threading my Way* (London, 1874).

—— *Moral Physiology* (London, 1877).

OWEN, ROBERT, and BRINDLEY, J., *What is Socialism? and what would be its Practical Effects upon Society? A Correct Report of the Public Discussion between Robert Owen and Mr J. Brindley held in Bristol* (London, 1841*a*).

—— —— *Public Discussion between John Brindley and Robert Owen, on the Question 'What is Socialism, and what would be its Practical Effects upon Society?'* (Chester, 1841*b*).

— and LEGG, WILLIAM, *Report of the Discussion between Robert Owen Esq and the Rev. Wm. Legg which Took Place in the Town Hall, Reading March 5 & 6, 1839, on Mr Owen's New Views of Society* (London, 1839).

'Oxoniensis', *Early Marriage and Late Parentage* (London, 1883).

PALEY, MORTON, '"A New Heaven is Begun": William Blake and Swedenborgianism', *Blake: An Illustrated Quarterly*, 13 (1979), 64–90.

PALEY, WILLIAM, *The Principles of Moral and Political Philosophy* (London, 1785).

PALMER, J. H., *Individual, Family and National Poverty* (London, 1878).

PANKHURST, RICHARD K.P., 'Anna Wheeler: A Pioneer Socialist and Feminist', *Political Quarterly*, 25 (1954), 132–44.

—— *The Saint Simonians Mill and Carlyle* (London, 1957).

PARSONS, BENJAMIN, *The Mental and Moral Dignity of Woman* (London, 1842).

PARSONS, GERALD, 'A Question of Meaning: Religion and Working-Class Life', in Gerald Parsons, ed., *Religion in Victorian Britain* (Manchester, 1988), ii. 63–87.

PATMORE, COVENTRY, 'Swedenborgiana', *National Review*, 6 (1858), 336–59.

PEARL, CYRIL, *The Girl with the Swansdown Seat* (London, 1955).

PECKHAM, MORSE, 'Victorian Counterculture', *Victorian Studies*, 18 (1975), 257–76.

PEEL, JOHN, 'Birth Control and the British Working-Class Movement', *Bulletin of the Society for the Study of Labour History*, 7 (1963), 16–22.

—— 'The Manufacturing and Retailing of Contraceptives in England', *Population Studies* 17 (1964), 113–25.

'Philanthropus', *The Institution of Marriage in the United Kingdom* (London, 1879).

PIKE, J. G., *Swedenborgianism in its True Colours* (Accrington, 1844).

PLACE, FRANCIS, *The Autobiography of Francis Place*, ed. Mary Thale (Cambridge, 1972).

PLUMPTRE, JAMES, *Forbidding to Marry* (Cambridge, 1812).

PODMORE, FRANK, *Robert Owen* (London, 1906).

POPE, NORRIS, *Dickens and Charity* (London, 1978).

POULSON, EDWARD, *A History of the Elements of Social Science* (London, 1892).

Poverty and Population (Nottingham, 1877).

PRENTICE, ARCHIBALD, *Historical Sketches and Personal Recollections of Manchester* (London, 1851).

PRICE, D. J. W., *A History of St David's University College Lampeter* (Cardiff, 1977).

PRIESTLEY, JOSEPH, *Considerations for the Use of Young Men* (New edn. London, 1791a).

—— *Letters to the Members of the New Jerusalem Church* (Birmingham, 1791b).

PRINCE, HENRY JAMES, *The Strength of Jesus to Perform Duty* (Bath, 1842).

—— *The Little Book Open* (London, 1856–9).

—— *The Testimony of the Two Anointed Ones* (London, 1858).

—— *Br. Prince's Journal* (London, 1859).

—— *The Counsel of God in Judgement* (1st edn. London, 1887; 2nd edn. London, 1900).

PROCHASKA, F. K., *Women and Philanthropy in Nineteenth-Century England* (Oxford, 1980).

PROTHERO, IORWETH, *Artisans and Politics in Early Nineteenth-Century London* (London, 1979).

PROUD, J., *A Candid and Impartial Reply to the Rev. Dr Priestley's Letters* (London, 1791).

—— *The Aged Minister's Last Legacy to the New Church* (Birmingham, 1818).

'Publicola', *Socialism Exposed* (London, 1844).

QUINLAN, MAURICE J., *Victorian Prelude* (London, 1941).

RADZINOWICZ, LEON, *A History of the Criminal Law and its Administration from 1750*, ii, iii (London, 1956); v, with R. Hood (London, 1986).

RATCLIFFE, B. M., and CHALONER, W. H., *A French Sociologist Looks at Britain* (Manchester, 1977).

RAZZELL, P. E., and WAINWRIGHT, R. W., *The Victorian Working Class* (London, 1973).

REES, ARTHUR AUGUSTUS, *The Rise and Progress of the Heresy of the Rev. H. J. Prince* (Weymouth, 1846).

Reformatory and Refuge Union, *Reports* (1866–77).

REID, J. C., *The Mind and Art of Coventry Patmore* (London, 1957).

RENDLE, T. McDONALD, *Swings and Roundabouts* (London, 1919).

Rescue Society, *Eighteenth Annual Report* (1871).

RITCHIE, J. EWING, *The Night Side of London* (London, 1857).

—— *The Religious Life of London* (London, 1870).

ROBERT, JEAN FRANÇOIS ('M.R.'), *Souvenirs d'Angleterre* (Lille, 1841).

ROBSON, ANN, 'The Noble Sphere of Feminism', *Victorian Periodicals Newsletter/Review*, 20 (1987), 102–7.

ROCHESTER, THOMAS (ed.), *Five Prize Essays on Man's Duty to God* (London, 1858).

ROMERO, PATRICIA W., *E. Sylvia Pankhurst* (New Haven, Conn., 1987).

ROSS, WILLIAM STEWART, ('Saladin'), *Historical Pamphlets* (London, 1887).

ROSSI, ALICE S., 'Sentiment and Intellect: The Story of John Stuart Mill and Harriet Taylor Mill', in Alice S. Rossi, ed., *John Stuart Mill and Harriet Taylor Mill, Essays on Sex Equality* (Chicago, 1970), 3–63.

ROUSSEAU, JEAN-JACQUES, *Emile* (London, 1974).

ROUTH, C. H. F., *The Moral and Physical Evils Likely to Follow if Practices Intended to Act as Checks to Population be not Strongly Discouraged and Condemned* (London, 1879).

ROVER, CONSTANCE, *Love, Morals and the Feminists* (London, 1970).

ROYLE, EDWARD, *Victorian Infidels* (Manchester, 1974).

RUBICHON, MAURICE, *De l'Angleterre*, ii (Paris, 1819).

RUDALL, JOHN, *A Memoir of the Rev. James Crabb* (London, 1854).

RUSHTON, ADAM, *My Life* (Manchester, 1909).

RUSSELL, BERTRAND, and PATRICIA, *The Amberley Papers* (London, 1937).

RUSSELL, G. W. E., *Collections and Recollections* (London, 1899), *Second Series* (London, 1908).

RYAN, MICHAEL, *Lectures on Population, Marriage, and Divorce* (London, 1831).

—— *The Philosophy of Marriage* (London, 1837).

—— *Prostitution in London* (London, 1839).

St David's College, Lampeter: Its Assailants and Defenders (London, 1851).

Saint Mary-le-bone Female Penitentiary Society, *Third Report* (London, 1841).

SALA, GEORGE, *Twice Round the Clock* (London, 1859).

SARGANT, WILLIAM LUCAS, *Economy of the Labouring Classes* (London, 1857).

SAVAGE, GAIL L., '"Intended only for the Husband": Gender, Class, and the Provision for Divorce in England, 1858–1868', in Kristine Otteson Garrigan, ed., *Victorian Scandals* (Athens, Oh., 1992), 11–42.

SAVILLE, JOHN, 'J. E. Smith and the Owenite Movement, 1833–1834', in Sidney Pollard and John Salt, eds., *Robert Owen Prophet of the Poor* (London, 1971), 115–44.

SAVILLE, JOHN, 'Robert Owen on the Family and the Marriage System of the Old Immoral World', in Maurice Cornforth, ed., *Rebels and their Causes* (London, 1978), 107–21.

The Science of Love (London, 1792).

SCORESBY, WILLIAM, *Lectures on Socialism* (London, 1840).

SCOTT, CLEMENT, 'The Old Coal Hole', *Daily Telegraph*, 20 Nov. 1896, p. 5.

—— *The Wheel of Life* (London, 1897).

—— and HOWARD, CECIL (eds.), *The Life and Reminiscences of E. L. Blanchard* (London, 1891).

SCOTT, PATRICK, 'The Business of Belief: The Emergence of "Religious" Publishing', *Studies in Church History*, 10 (1973), 213–24.

SCUDAMORE, W. E., *An Account of the Penitentiary at Shipmeadow* (London, 1857).

SEARLE, G. R., 'Socialism and Malthusianism in Late Victorian and Edwardian Britain', in J. Dupaquier, A. Fauve-Chamoux, and E. Grebenik, eds., *Malthus Past and Present* (London, 1983), 341–56.

SELLERS, EDITH, 'The Founder of a Peculiar Sect', *Newbery House Magazine*, 5 (1891), 586–95.

SHANLEY, MARY LYNDON, *Feminism, Marriage, and the Law in Victorian England, 1850–1895* (Princeton, NJ, 1989).

SHANNON, RICHARD Q., 'Public Amusements: Pretensions of the Evangelical Class', *Edinburgh Review*, 54 (1831), 100–14.

SHAW, CHARLES, *When I was a Child* (Firle, 1977).

SHAW, GEORGE BERNARD, *Sixteen Self Sketches* (London, 1949).

SHIPLEY, STAN, *Club Life and Socialism in Mid-Victorian London* (History Workshop Pamphlets, 5; Oxford, 1971).

'Simplex', *An Inquiry into the Constitution, Government, and Practices of the Churches of Christ* (Edinburgh, 1808).

SIMS, GEORGE R., *My Life* (London, 1917).

SKENE, FELICIA M., *Penitentiaries and Reformatories* (Edinburgh, 1865).

SKIDMORE, W. J., *A Fifth Letter to the Followers of Emmanuel Swedenborg* (Haslingden, 1844).

SMILES, SAMUEL, 'The Condition of Factory Women: What is doing for them?', *People's Journal*, 2 (1846), 358–60; 3 (1847), 52–4, 143–4.

SMITH, AGNES M., 'The First Mormon Mission to Britain', *History Today* (July 1987), 24–31.

SMITH, CHARLES MANBY, *A Working-Man's Way in the World* (London, 1853).

SMITH, F. B., 'The Atheist Mission 1840–1900', in Robert, Robson, ed., *Ideas and Institutions of Victorian Britain* (London, 1967), 115–42.

—— 'Sexuality in Britain, 1800–1900', *University of Newcastle Historical Journal*, 2 (1974), 19–32.

SMITH, FREDERIC, *An Appeal to the Virtue and Good Sense of the Inhabitants of St Martin in the Fields and Other Parishes in Westminster* (London, 1809).

SMITH, JAMES ELISHAMA, *Lecture on a Christian Community* (London, 1833*a*).

—— *The Antichrist* (London, 1833*b*).

—— *The Little Book* (London, 1840).

SMITH, SIDNEY, *The Works of the Rev. Sidney Smith* (London, 1859).

SMITH, W. ANDERSON, *'Shepherd' Smith the Universalist* (London, 1892).

SMITH, WARREN SYLVESTER, *The London Heretics 1870–1914* (London, 1967).

SMITH, WILLIAM HAWKES, *Letters on Social Science* (Birmingham, 1839).

SNELL, HENRY, *Men, Movements, and Myself* (London, 1936).

Social Science: Being Selections from John Cassell's Prize Essays (London, 1861).

Society for the Suppression of Vice, *Address to the Public* (1803).

—— *Occasional Reports* (1810–68).

SOUTHWELL, CHARLES, *An Essay on Marriage* (London, 1840*a*).

—— *Socialism Made Easy* (London, 1840*b*).

—— *Confessions of a Freethinker* (London, 1845).

SPENCE, WILLIAM, *Essays in Divinity and Physic* (London, 1792).

'S.T.', 'An Address to the Guardian Society', *Pamphleteer*, 11 (1818), 226–52.

STANDRING, GEORGE, 'The Author of "The Elements"', *Freethinker*, 18 Dec. 1904, p. 813.

Statement of the Origin, Proceedings and Results of the Midnight Meetings for the Recovery of Fallen Women (London, 1860).

STEWART, ALLAN, *The Abominations of the Wroeites* (Melbourne, 1864).

STEWART, DUGALD, *The Collected Works of Dugald Stewart*, vii, ed. William Hamilton (Edinburgh, 1855).

STONE, LAWRENCE, *Road to Divorce* (Oxford, 1990).

STOPES, MARIE CARMICHAEL, *The Early Days of Birth Control* (London, 1922).

STRUTT, ELIZABETH, *The Feminine Soul: Its Nature and Attributes* (London, 1857).

STURMER, FREDERICK, *Socialism, its Immoral Tendency* (London, 1840).

SULLIVAN, ALVIN, *British Literary Magazines: The Romantic Age 1789–1836* (Westport, 1983).

SWAN, GUIDA (ed.), *The Journals of Two Poor Dissenters 1786–1880* (London, 1970).

SWEDENBORG, EMMANUEL, *The Delights of Wisdom Concerning Conjugial Love* (London, 1794).

TAIT, WILLIAM, *Magdalenism* (Edinburgh, 1840).

TALBOT, JAMES BEARD, *The Miseries of Prostitution* (London, 1844).

TAYLER, WILLIAM, *Diary of William Tayler, Footman*, ed. Dorothy Wise (London, 1962).

TAYLOR, ANNE, *Annie Besant* (Oxford, 1992).

TAYLOR, BARBARA, *Eve and the New Jerusalem* (London, 1983).

TAYLOR, SHEPHARD, T., *The Diary of a Medical Student during the Mid-Victorian Period 1860–1864* (Norwich, 1927).

TAYLOR, WILLIAM COOKE, 'The Employment of Married Women in Manufacture', *Transactions of the National Association for the Promotion of Social Science 1873* (London, 1874), 605–13.

TAYLOR, WILLIAM J., *The Story of the Homes* (London, 1907).

TEXIER, EDMOND, *Lettres sur l'Angleterre* (Paris, 1851).

THOLFSEN, TRYGVE R., *Working-Class Radicalism in Mid-Victorian England* (London, 1976).

THOMAS, EDWARD W., *Twenty-Five Years' Labour among the Friendless and Fallen* (London, 1879).

THOMAS, FRANCES, *Christina Rossetti* (Hanley Swan, Hereford and Worcester, 1992).

THOMAS, GEORGE ROBINSON, and STARKEY, SAMUEL, *The Revelation of Jesus Christ as the Son of Man* (London, 1862).

THOMASON, DENNY R., *Fashionable Amusements* (London, 1827).

THOMPSON, E. P., *The Making of the English Working Class* (Harmondsworth, 1968).

THOMPSON, WILLIAM, *Appeal of One Half of the Human Race, Women, against the Pretensions of the Other Half, Men, to Retain them in Political, and thence in Civil and Domestic, Slavery* (London, 1825).

—— *Practical Directions for the Speedy and Economic Establishment of Communities* (London, 1830).

THORNE, WILL, *My Life's Battles* (London, 1925).

TILLETT, BEN, *Memories and Reflections* (London, 1931).

TRAILL, H. D., 'The Abdication of Mrs Grundy', *National Review*, 17 (1891), 12–24.

TRELAWNEY, E. J., *Adventures of a Younger Son* (London, 1831).

TRIBE, DAVID, *President Charles Bradlaugh MP* (London, 1971).

TRISTAN, FLORA, *London Journal* (London, 1980).

TROBRIDGE, GEORGE, 'Coventry Patmore and Swedenborg', *Westminster Review*, 165 (1906), 76–91.

TROUP, G., and JONES, LLOYD, *Report of the Discussion betwixt Mr Troup, Editor of the Montrose Review, on the Part of the Philalethean Society, and Mr Lloyd Jones, of Glasgow, on the Part of the Socialists* (Dundee, 1839).

TUCKERMAN, HENRY T., *A Month in England* (London, 1854).

TURNER, GEORGE, *The Assurance of the Kingdom* (London, 1819a).

—— *Wonderful Prophecies by George Turner*, pt. ii (London, 1819b).

—— *The Marriage of the Lamb, Revealed to George Turner* (London, 1820a).

—— *The Rich Treasure of the Kingdom, Revealed to George Turner* (London, 1820b).

TYRRELL, ALEXANDER, 'Class Consciousness in Early Victorian Britain: Samuel Smiles, Leeds Politics, and the Self-Help Creed', *Journal of British Studies*, 9 (1970), 102–25.

VANCE, NORMAN, *The Sinews of the Spirit* (Cambridge, 1985).

VANDERKISTE, R. W., *Notes and Narratives of a Six Years' Mission* (London, 1852).

VINCENT, DAVID MICHAEL, *Bread, Knowledge and Freedom* (London, 1981).

VIZETELLY, HENRY, *Glances Back through Seventy Years* (London, 1893).

VOGELER, MARTHA S., *Frederick Harrison: The Vocations of a Positivist* (Oxford, 1984).

VOLNEY, COMTE DE, *The Law of Nature* (n.pl., ?1846).

WADE, JOHN, *Women, Past and Present* (London, 1859).

WAGNER, LEOPOLD, *More London Inns and Taverns* (London, 1925).

WAKEFIELD, EDWARD GIBBON, *England and America* (London, 1833).

WALKER, ALEXANDER, *Woman Physiologically Considered, as to Mind, Morals, Marriage, Matrimonial Slavery, Infidelity and Divorce* (2nd edn. London, 1840).

WALKOWITZ, JUDITH R., *Prostitution and Victorian Society* (Cambridge, 1980).

—— 'Male Vice and Feminist Virtue: Feminism and the Politics of Prostitution in Nineteenth-Century Britain', *History Workshop*, 13 (1982), 79–89.

WEBSTER, AUGUSTA, *A Housewife's Opinions* (London, 1879).

WHALLEY, R., *A Philosophical Refutation of the Theories of Robert Owen and his Followers* (Manchester, 1840).

WIENER, JOEL H., 'The Journalism of Richard Carlile', *Victorian Periodicals Newsletter*, 9 (1976), 75–80.

WILKINSON, CLEMENT JOHN, *James John Garth Wilkinson* (London, 1911).

WILLARD, EMMA, *Journal and Letters, from France and Great Britain* (Troy, 1833).

WILLIAMS-HOGAN, JANE, *A New Church in a Disenchanted World*, Univ. of Pennsylvania, Ph.D. thesis (1985).

WILSON, JOHN, *Autobiography* (Durham, 1909).

WILSON, JOHN, 'Mr Sadler and the Edinburgh Reviewer . . .', *Blackwoods*, 29 (1831), 392–428.

WOLLSTONECRAFT, MARY, *Vindication of the Rights of Woman* (Harmondsworth, 1975).

WOODMAN, WOODVILLE, *Swedenborg's Conjugial Love* (Leamington Spa, 1858).

—— *Swedenborg's Doctrine of Marriage and its Opposites* (London, 1865).

WRIGHT, HENRY G., *Marriage and its Sanctions* (Cheltenham, 1840).

WRIGHT, THOMAS, *The Great Unwashed* (London, 1868).

YATES, EDMUND HODGSON, *Edmund Yates: His Recollections and Experiences* (London, 1884).

YOUNG, G. M., *Victorian England Portrait of an Age* (London, 1960).

PERIODICALS

Antidote, Antidote to Mormonism and Infidel Error, Anti-Socialist Gazette, Athenaeum, Aurora, Barker's Review, Biograph and Review, Bridgwater Times, British and Foreign Medical Review, British and Foreign Medico-Chirurgical Review, British Controversialist, British Medical Journal, British Quarterly, Bury Times, Christian Lady's Magazine, Christian Observer, Christian Reformer, Christian Review and Clerical Magazine, Contemporary Review, Co-operative News, Cornhill, Crisis, Crown, Daily Telegraph, Dickensian, Examiner, Exquisite, Fast Life, Female's Advocate, Female's Friend, Fortnightly Review, Frasers Magazine, Friend of the Fallen, Gauntlet, Household Words, Illustrated London Life, Illustrated London News, Intellectual Repository, Investigator, Isis, Lancet, Livesey's Moral Reformer, London Gazette, Magazine of Knowledge, Magdalen's Friend, Malthusian, Man, Medical Press and Circular, Medical Times and Gazette, Metropolitan, Monthly Magazine, Monthly Repository, Moral Reformer, Morning Chronicle, National Reformer, National Review, National Secular Society's Almanack, New Age, New Church Repository, New Moral World, Newgate Magazine, Pathfinder, Paul Pry, People's Journal, People's Paper, Political Economist, Present Day, Public Health, Quarterly Review, Reasoner, Reformatory and Refuge Union Journal, Robert Owen's Journal, St Paul's Magazine, Sanitary Review and Journal of Public Health, Saturday Review, Scourge, Secular World, Shepherd, Social Pioneer, Taits, Taunton Courier and Western Advertiser, Temple Bar, Times, Town, Weekly Dispatch, Westminster Review.

Index

INDEX